United States Nuclear Regulatory Commission

Protecting People and the Environment

NUREG-1650
Revision 4

The United States of America National Report for the 2012 Convention on Nuclear Safety Extraordinary Meeting

Office of Nuclear Reactor Regulation

AVAILABILITY OF REFERENCE MATERIALS
IN NRC PUBLICATIONS

NRC Reference Material

As of November 1999, you may electronically access NUREG-series publications and other NRC records at NRC's Public Electronic Reading Room at http://www.nrc.gov/reading-rm.html.
Publicly released records include, to name a few, NUREG-series publications; *Federal Register* notices; applicant, licensee, and vendor documents and correspondence; NRC correspondence and internal memoranda; bulletins and information notices; inspection and investigative reports; licensee event reports; and Commission papers and their attachments.

NRC publications in the NUREG series, NRC regulations, and *Title 10, Energy*, in the Code of *Federal Regulations* may also be purchased from one of these two sources.
1. The Superintendent of Documents
   U.S. Government Printing Office
   Mail Stop SSOP
   Washington, DC 20402–0001
   Internet: bookstore.gpo.gov
   Telephone: 202-512-1800
   Fax: 202-512-2250
2. The National Technical Information Service
   Springfield, VA 22161–0002
   www.ntis.gov
   1–800–553–6847 or, locally, 703–605–6000

A single copy of each NRC draft report for comment is available free, to the extent of supply, upon written request as follows:
Address: U.S. Nuclear Regulatory Commission
          Office of Administration
          Publications Branch
          Washington, DC 20555-0001
E-mail: DISTRIBUTION.SERVICES@NRC.GOV
Facsimile: 301–415–2289

Some publications in the NUREG series that are posted at NRC's Web site address
http://www.nrc.gov/reading-rm/doc-collections/nuregs
are updated periodically and may differ from the last printed version. Although references to material found on a Web site bear the date the material was accessed, the material available on the date cited may subsequently be removed from the site.

Non-NRC Reference Material

Documents available from public and special technical libraries include all open literature items, such as books, journal articles, and transactions, *Federal Register* notices, Federal and State legislation, and congressional reports. Such documents as theses, dissertations, foreign reports and translations, and non-NRC conference proceedings may be purchased from their sponsoring organization.

Copies of industry codes and standards used in a substantive manner in the NRC regulatory process are maintained at—
          The NRC Technical Library
          Two White Flint North
          11545 Rockville Pike
          Rockville, MD 20852–2738

These standards are available in the library for reference use by the public. Codes and standards are usually copyrighted and may be purchased from the originating organization or, if they are American National Standards, from—
          American National Standards Institute
          11 West 42nd Street
          New York, NY 10036–8002
          www.ansi.org
          212–642–4900

Legally binding regulatory requirements are stated only in laws; NRC regulations; licenses, including technical specifications; or orders, not in NUREG-series publications. The views expressed in contractor-prepared publications in this series are not necessarily those of the NRC.

The NUREG series comprises (1) technical and administrative reports and books prepared by the staff (NUREG–XXXX) or agency contractors (NUREG/CR–XXXX), (2) proceedings of conferences (NUREG/CP–XXXX), (3) reports resulting from international agreements (NUREG/IA–XXXX), (4) brochures (NUREG/BR–XXXX), and (5) compilations of legal decisions and orders of the Commission and Atomic and Safety Licensing Boards and of Directors' decisions under Section 2.206 of NRC's regulations (NUREG–0750).

United States Nuclear Regulatory Commission

Protecting People and the Environment

NUREG-1650
Revision 4

The United States of America National Report for the 2012 Convention on Nuclear Safety Extraordinary Meeting

Manuscript Completed: May 2012
Date Published: July 2012

Prepared by
U.S. Nuclear Regulatory Commission (NRC)
Institute of Nuclear Power Operations (INPO)

ABSTRACT

The United States (U.S.) Nuclear Regulatory Commission (NRC), in coordination with the U.S. Department of State, U.S. Department of Energy, and the Institute of Nuclear Power Operations, prepared this report, "The United States of America National Report for the 2012 Convention on Nuclear Safety Extraordinary Meeting," and will submit it for peer review at the 2012 Convention on Nuclear Safety (CNS) Extraordinary Meeting to be held at the International Atomic Energy Agency in Vienna, Austria, in August 2012. This report addresses the actions taken by the U.S. to improve nuclear safety in response to the March 11, 2011, accident at the Fukushima Daiichi nuclear power plant in Japan. It demonstrates how the U.S. contributes to achieving and maintaining a high level of nuclear safety worldwide by enhancing national measures and international cooperation and by meeting the obligations of all the articles established by CNS. It describes how the U.S. addressed six topics in relation to the Fukushima accident: (1) external events, (2) design issues, (3) severe accident management and recovery, (4) national organizations, (5) emergency preparedness and response and post-accident management, and (6) international cooperation. Similar to the U.S. National Report for the fifth CNS issued in 2010, this report includes a section developed by the Institute of Nuclear Power Operations describing work done by the U.S. nuclear industry in response to the Fukushima accident.

CONTENTS

EXECUTIVE SUMMARY

On March 11, 2011, the Tohoku-Taiheiyou-Oki earthquake and subsequent tsunami hit Japan, devastating the area and affecting numerous nuclear power plants, including Fukushima Daiichi.

At the fifth review meeting of the Convention on Nuclear Safety (CNS), held April 4 – 15, 2011, in Vienna, Austria, the contracting parties agreed to analyze the relevant issues of the Fukushima accident during an extraordinary meeting. This meeting is scheduled to take place August 27 – 31, 2012, in Vienna, Austria. The objectives of this CNS extraordinary meeting are to enhance safety through reviewing and sharing lessons learned and actions taken by contracting parties in response to the events of Fukushima and to review the effectiveness and continued stability of the provisions of the CNS.

Consistent with this commitment, the United States (U.S.) developed this national report detailing the response of the U.S. Nuclear Regulatory Commission and the U.S. nuclear industry to the safety issues raised by the Fukushima accident. The following topics are addressed:

(1) external events
(2) design issues
(3) severe accident management and recovery
(4) national organizations
(5) emergency preparedness and response and post-accident management
(6) international cooperation

Part 1 of the report describes how the U.S. Nuclear Regulatory Commission is addressing these six topic areas. Because the prime responsibility for the safety of a nuclear installation rests with the license holder, Part 2 of the report explains how the nuclear industry is addressing the six topics described above. Part 2 was developed by the Institute of Nuclear Power Operations in cooperation with other entities including the Nuclear Energy Institute. Each topic addresses actions taken, actions planned, schedules, and results. The report also contains a table summarizing all the actions and activities discussed herein. This matrix is located in Appendix A.

ACKNOWLEDGMENTS

Contributors to this report include the following technical and regulatory experts from the U.S. Nuclear Regulatory Commission:

Astwood, Heather
Brown, David
Couret, Ivonne
Decker, David
Dudek, Michael
Foster, Jack
Garmon, David
Harrington, Holly
Henderson, Karen
Hopkins, Jon
Kauffman, John
Kratchman, Jessica
Laura, Richard
Miller, Barry
Mitchell, Matthew
Norton, Charles
Pascarelli, Robert
Quiñones-Navarro, Lauren
Rodriguez, Veronica
Schwartzman, Jennifer
Shane, Raeann
Skeen, David
Taylor, Robert
Thorp, John
Witt, Kevin
Wittick, Susan

Contributors to this report include the following experts from the Institute of Nuclear Power Operations:

Nielsen, Frederick
Shorey, Larry
Webster, William

ABBREVIATIONS AND ACRONYMS

AC	alternating current
ACIRS	Advisory Committee for the International Reporting System for Operating Experience
ADAMS	Agencywide Documents Access and Management System
ANPR	advance notice of proposed rulemaking
AOP	abnormal operating procedure
BWR	boiling water reactor
CEUS	Central and Eastern United States
CFR	*Code of Federal Regulations*
CNRA	Committee on Nuclear Regulatory Activities
CNS	Convention on Nuclear Safety
CNSC	Canadian Nuclear Safety Commission
CNSNS	Comisión Nacional de Seguridad Nuclear y Salvaguardias of Mexico
COL	combined license
DC	direct current
DHS	U.S. Department of Homeland Security
DOD	U.S. Department of Defense
DOE	U.S. Department of Energy
EDG	emergency diesel generator
EDMG	extensive damage mitigation guidelines
EDO	Executive Director for Operations
EOP	emergency operating procedure
EP	emergency preparedness
EPRI	Electric Power Research Institute
ERDS	Emergency Response Data System
EPA	Environmental Protection Agency
ESP	early site permit
FEMA	Federal Emergency Management Agency
FLEX	Diverse and Flexible Mitigation Capability
GDC	general design criterion
GIP	Generic Issues Program
GL	generic letter
HOC	Headquarters operations center
IAEA	International Atomic Energy Agency
IN	information notice
INES	International Nuclear and Radiological Event Scale
INPO	Institute of Nuclear Power Operations
IPEEE	individual plant examinations for external events
IRS	International Reporting System for Operating Experience

JNES	Japan Nuclear Energy Safety Organization
JLD	Japan Lessons-Learned Project Directorate
KI	potassium iodide
LOOP	loss of offsite power
MDEP	Multinational Design Evaluation Program
NEA	Nuclear Energy Agency
NEI	Nuclear Energy Institute
NOAA	National Oceanic and Atmospheric Administration
NPP	nuclear power plant
NRC	U.S. Nuclear Regulatory Commission
NRR	Office of Nuclear Reactor Regulations
NTTF	Near-Term Task Force
OECD	Organisation for Economic Co-operation and Development
OCA	Office of Congressional Affairs
OPA	Office of Public Affairs
PRA	probabilistic risk assessment
RFI	request for information
RG	regulatory guide
SAMG	severe accident management guidelines
SBO	station blackout
SECY	Office of the Secretary of the Commission
SFP	spent fuel pool
SRM	staff requirements memorandum
SSC	structure, system, and component
TEPCO	Tokyo Electric Power Company
TI	temporary instruction
TMI	Three Mile Island Nuclear Generating Station
UHS	ultimate heat sink
WANO	World Association of Nuclear Operators
WGOE	Working Group for Operating Experience

DEFINITIONS

Advanced Notice of Proposed Rulemaking (ANPR) – An ANPR is a vehicle for soliciting external stakeholder feedback concerning a potential regulatory action. Its use can vary depending on the circumstances associated with the regulatory issue under consideration, including soliciting feedback to determine whether an issue warrants regulatory action, collecting information that can be used to complete a regulatory basis for a potential regulatory action, or obtaining comments on proposed regulatory approaches. Issuance of ANPR does not, however, commit the NRC to a rulemaking.

Beyond-design-basis events – events that are possible but were not fully considered in the design process because they were judged to be too unlikely.

***Code of Federal Regulations* (CFR)** – the codification of the general and permanent rules published in the Federal Register by the executive departments and agencies of the U.S. Government.

Combined license (COL) – an NRC-issued license that authorizes a licensee to construct and (with certain specified conditions) operate a nuclear power plant (NPP) at a specific site, in accordance with established laws and regulations. A COL is valid for 40 years (with the possibility of a 20-year renewal).

Diverse and Flexible Mitigation Capability (FLEX) – a U.S. industry strategy being developed to address the impact of external events. The strategy includes the use of portable contingency equipment positioned at diverse on-site locations capable of being connected to maintain or restore core cooling, containment and spent fuel pool cooling capabilities

Early Site Permit (ESP) – an ESP is an NRC approval for a site (or sites) for one or more nuclear power facilities. This includes a partial constructions permit. The approval of an ESP provides an applicant with an opportunity to hold a site for up to 20 years, reduces licensing uncertainty, and resolves siting issues before construction begins.

***Federal Register* notice** – publication of a document in the *Federal Register*, the official daily publication for rules, proposed rules, and notices of U.S. Federal agencies and organizations, as well as executive orders and other presidential documents.

Generic letter (GL) – a letter that addresses either (1) a technical issue, emergent or routine, that has arisen between the NRC staff and the nuclear industry, or (2) a compliance matter that the NRC staff has concluded carries significant risk and should therefore be brought immediately to the attention of the nuclear industry. A GL may request information from, or request action, by the addressees on matters of safety, safeguards, or environmental significance. A GL may request new or revised commitments based on analyses performed and proposed corrective actions but may not require commitments.

Hostile-Action-Based Drill – an exercise that simulates an act toward an NPP or its personnel that includes the use of violent force to destroy equipment, take hostages, or intimidate the licensee to achieve an end.

Information Notice (IN) – a communication used by the NRC to inform the U.S. nuclear industry of recently identified, significant operating experience. The nuclear industry is expected to review the information for applicability to its facilities or operations and consider actions, as appropriate, to avoid similar problems. An IN neither conveys nor implies new requirements and interpretations, nor does it request information or action.

National Response Framework – the guiding principles that enable all response partners to prepare and provide a unified national response to disasters and emergencies - from the smallest incident to the largest catastrophe. The framework establishes a comprehensive, national, all-hazards approach to domestic incident response.

Nuclear/Radiological Incident Annex (NRIA) to the National Response Framework – the set of policies, situations, concepts of operations, and responsibilities of Federal departments and agencies that governs the immediate response and short-term recovery activities for incidents involving the release of radioactive materials.

Probabilistic Risk Assessment (PRA) – a systematic analysis tool that consists of specific technical elements that provide both *qualitative insights* and a *quantitative assessment of risk* by addressing the following questions known as the "risk triplet": (1) What can go wrong?, (2) How likely is it?, and (3) What are the consequences? Modern PRAs also have incorporated uncertainty analyses to address a fourth question: How confident are we in our answers to these three questions? In this way, PRAs improve NPP safety by identifying, prioritizing, and mitigating significant contributors to risk. Three levels of PRA address different aspects of accidents:

- *Level 1 PRA* uses event and fault trees to model the responses of NPPs and their operators to initiating events that challenge plant operation. These models identify accident sequences that result in damage to the reactor core. The estimated frequencies for all accident sequences that cause core damage are summed to calculate the NPP's total core damage frequency.

- *Level 2 PRA* models and analyzes the progression of "severe accidents" - Level 1 PRA accident sequences that result in reactor core damage - by considering how the reactor coolant and other relevant systems and the containment respond to the accident. This analysis is based on both the initial status of structures and systems and their ability to withstand the harsh accident environment. Once the system and containment response is characterized, the PRA can determine the frequency, type, amount, timing, and energy content of the radioactivity released to the environment, also known as "source term characteristics".

- *Level 3 PRA* models the release and transport of radioactive material in a severe accident and estimates the health and economic impact in terms of the following measures of offsite consequences: (1) early fatalities and injuries and latent cancer fatalities resulting from the radiation doses to the surrounding population and (2) economic costs associated with evacuation, relocation, property loss, and decontamination. Offsite consequences are estimated using the Level 2 PRA source term characteristics and several other factors that affect the transport and impact of the radioactive material. These factors include meteorology, demographics, emergency response, and land use. By combining the results of the Level 1 and Level 2 PRAs with the results of this consequence analysis, the Level 3 PRA estimates the NPP's integrated risk (likelihood × consequences) to the public.

Order – A written NRC directive used to modify, suspend, or revoke a license or to require specific actions by licensee or other persons. Orders can also be used to impose civil penalties.

Regulatory guide (RG) – a document that offers guidance to licensees and applicants on implementing specific parts of the NRC's regulations, techniques used by the NRC staff to evaluate specific problems or postulated accidents, and data needed the staff needs for its review of applications for permits or licenses.

Request for information (RFI) – a written request from the NRC to a licensee that seeks information about a potential safety issue. The RFI typically is used to determine whether to modify, suspend or revoked a license. The NRC must justify each request to ensure it addresses a potentially significant safety issue. These requests are issued as letters under the authority of Title 10 of the *Code of Federal Regulations* Section 50.54(f).

Rulemaking – the process by which the NRC develops regulations.

SECY paper – documents on policy, rulemaking, and adjudicatory matters, as well as general information that the NRC staff provides to the Commission for consideration.

Safe-shutdown Earthquake – earthquake ground shaking for which certain structures, systems, and components are designed to remain functional.

Seiches – an occasional and sudden oscillation of a body of water (e.g., lake, bay, or estuary), caused by events such as wind, earthquakes, or changes in barometric pressure, that produces fluctuations in the water level.

Staff requirements memorandum (SRM) – a concise statement of the Commission's decision about a SECY paper's recommendation that notes any approved modifications to the recommendation and clearly states any additional requirements the staff must perform, along with due dates.

Stakeholder – a person or group that has an investment, share, or interest in the NPP industry, including private citizens, licensees, and governmental and nongovernmental organizations.

Station blackout (SBO) – the total loss of alternating current (AC) power at an NPP as a result of complete failure of both offsite and onsite AC power sources

Steering Committee – committee responsible for implementing the Near-Term Task Force recommendations. The committee is chaired by the Deputy Executive Director for Reactor Preparedness and includes Office Directors from many of the NRC's program offices and regions. The group is supported by the Japan Lessons-Learned Project Directorate.

Tier 1 actions – actions that are to begin without unnecessary delay. These actions are considered to have the greatest potential for safety improvement in the near-term.

Tier 2 actions – actions that cannot be initiated because of the need for further technical assessment and alignment, dependence on Tier 1 issues, or current lack of critical skill sets.

Tier 3 actions – actions that (1) require further study to support regulatory action, (2) depend on the completion of an associated short-term action to inform the longer-term action, (3) depend on the availability of critical skill sets, or (4) depend on the development of a logical, systematic, and coherent regulatory framework for adequate protection that appropriately balances defense-in-depth and risk considerations.

Ultimate Heat Sink (UHS) – consists of an assured supply of water that is credited for dissipating reactor decay heat and essential station heat loads after a normal reactor shutdown or a shutdown following an accident or transient, including a loss-of-coolant accident. The UHS includes the cooling media itself, the equipment needed to pump the cooling media, the heat exchangers and all possible combinations of these systems. Many commercial NPPs also rely upon the atmosphere for performing the UHS function to some extent in conjunction with the assured supply of cooling water, such as in the case of spray ponds and cooling towers; and passive reactor plant designs may rely more exclusively on the atmosphere for dissipating reactor decay heat immediately following plant transient and accident conditions.

PART 1

INTRODUCTION

The U.S. Nuclear Regulatory Commission (NRC) continuously oversees nuclear power plants to verify that they are being operated in accordance with the agency's rules and regulations. Immediately after the March 2011 earthquake in Japan, the NRC staffed its Headquarters operations center (HOC) and began communicating available information to stakeholders, including members of the public, the U.S. Congress, representatives from State Governments, other Federal agencies, and the nuclear industry, regarding the developments at the Fukushima Daiichi nuclear power plant. Staff from U.S. Government agencies, including the NRC, Department of Energy (DOE), and Department of Defense (DOD), was dispatched as an assistance team to support the U.S. Ambassador in Japan and the Japanese Government.

The NRC has the authority to take any action it deems necessary to protect public health and safety, and may demand immediate licensee response, including plant shutdowns. The NRC took immediate action following the Fukushima accident, in the form of temporary instructions (TIs), information notices (INs), bulletins, and inspections, to ensure that there were no immediate safety concerns at American facilities. Next, the NRC created the Near-Term Task Force (NTTF), composed of senior NRC staff and management, to systematically and methodically review the NRC's processes and regulations. The NRC tasked the NTTF with determining whether the NRC should make additional improvements to its regulatory system, and to make policy recommendations to the Commission. The NTTF issued its report, titled "Recommendations for Enhancing Reactor Safety in the 21st Century: The Near-Term Task Force Review of the Insights from the Fukushima Daiichi Accident," on July 12, 2011. The NTTF concluded that continued operation of U.S. nuclear plants and ongoing NRC licensing activities posed no imminent risk. The NTTF also concluded that enhancements to safety and emergency preparedness (EP) are warranted, and it made 12 overarching recommendations for Commission consideration.

Since then, the NRC has created a Steering Committee, which is chaired by the Deputy Executive Director for Reactor and Preparedness Programs and is composed of Office Directors from many of the NRC program offices and regions. The NRC also created a new organization, the Japan Lessons-Learned Project Directorate (JLD). The objective of the Steering Committee and the JLD is to perform a long-term review of the March 11, 2011, Japanese earthquake and tsunami, and to oversee implementation of Fukushima lessons learned for U.S. nuclear plants. All of the NRC's actions in response to the Fukushima Daiichi accident are being coordinated by these two groups. For example, the Steering Committee prioritized the NTTF's recommendations and other recommendations derived from interactions with international organizations; other Federal, State, and local agencies; NRC's Advisory Committee on Reactor Safeguards; the public; and other stakeholders. In some cases, recommendations from the Steering Committee modified or enhanced the NTTF's recommendations.

The NRC's prioritization scheme, SECY-11-0137, "Prioritization of Recommended Actions To Be Taken in Response to Fukushima Lessons Learned," dated October 3, 2011, describes three tiers of actions.

Tier 1 actions are to begin without unnecessary delay. To determine and recommend near-term regulatory actions that should be initiated without delay, the staff considered whether any of the recommendations identified an imminent hazard to public health and safety. The staff concluded that none of the recommendations rise to this level. The staff then identified, under the Tier 1 activities, a subset of actions that has the greatest potential for safety improvement in the near-term.

Tier 2 actions are those that cannot yet be initiated because of a need for further technical assessment and alignment, dependence on Tier 1 issues, or lack of availability of critical skill sets. Tier 3 actions are those that require further staff study to support regulatory action; need the result of an associated short-term action to inform the long-term action; depend on the availability of critical skill sets; or depend on the development of a logical, systematic, and coherent regulatory framework that balances defense-in-depth and risk considerations.

Part 1 of this report describes actions the NRC and the U.S. Government have taken, or plan to take to address lessons learned from the Fukushima accident as they pertain to U.S. nuclear power plants and international cooperation on nuclear safety. The topics addressed are (1) external events, (2) design issues, (3) severe accident management and recovery, (4) national organizations, (5) EP and response and post-accident management, and (6) international cooperation. Each topic covered in the report includes a discussion of the actions taken or planned, schedules for completion of ongoing and planned activities, and results of completed actions.

TOPIC 1. EXTERNAL EVENTS

1.1 Introduction

The NRC has long recognized that protection from natural phenomena is an important means to prevent core damage and ensure the integrity of containment and the spent fuel pool (SFP). The NRC established several requirements addressing natural phenomena in 1971 with General Design Criterion (GDC) 2, "Design Bases for Protection Against Natural Phenomena," of Appendix A, "General Design Criteria for Nuclear Power Plants," to Title 10 of the *Code of Federal Regulations* (10 CFR) Part 50, "Domestic Licensing of Production and Utilization Facilities."

GDC 2 requires, in part, that structures, systems, and components (SSCs) important to safety be designed to withstand the effects of natural phenomena such as floods, tsunami, and seiches without losing the capability to perform their safety functions. GDC 2 also requires that design bases for these SSCs reflect (1) appropriate consideration of the most severe of the natural phenomena that historically have been reported for the site and surrounding region, with sufficient margin for the limited accuracy and quantity of the historical data and the period of time in which the data have been accumulated; (2) appropriate combinations of the effects of normal and accident conditions with the effects of the natural phenomena; and (3) the importance of the safety functions to be performed.

Since the establishment of GDC 2, the NRC's requirements and guidance for protection from seismic events, floods, and other natural phenomena have continued to evolve. The NRC has developed new regulations, new and updated regulatory guidance, and several regulatory programs to enhance previously licensed reactors, including the following:

- Appendix A, "Seismic and Geologic Siting Criteria for Nuclear Power Plants," to 10 CFR Part 100, "Reactor Site Criteria"

- NUREG-0800, "Standard Review Plan for the Review of Safety Analysis Reports for Nuclear Power Plants: LWR [light water reactor] Edition"), and interim staff guidance in three areas related to protection from natural phenomena

- The systematic evaluation program established in 1977 to review the designs of older operating nuclear reactor plants to reconfirm and document their safety

- Issuing Generic Letter (GL) 87-02, "Verification of Seismic Adequacy of Mechanical and Electrical Equipment in Operating Reactors, Unresolved Safety Issue A-46," dated February 19, 1987, to address concerns related to seismic qualification of mechanical and electrical equipment in operating nuclear power plants (NPPs)

- Supplement 4 to GL 88-20, "Individual Plant Examination of External Events for Severe Accident Vulnerabilities, 10 CFR 50.54(f) (Subsection (f) of 10 CFR 50.54, "Conditions of Licenses")," dated June 28, 1991, requesting licensees to perform an individual plant examination of external events (IPEEE) to identify vulnerabilities

- Subpart B, "Evaluation Factors for Stationary Power Reactor Site Applications on or After January 10, 1997," to 10 CFR Part 100

- Appendix S, "Earthquake Engineering Criteria for Nuclear Power Plants," to 10 CFR Part 50

The staff also has published several regulatory guides (RGs) that address specific technical issues related to protection from natural phenomena, including the following:

- RG 1.29, "Seismic Design Classification," dated March 2007

- RG 1.59, "Design Basis Floods for Nuclear Power Plants," dated August 1977

- RG 1.60, "Design Response Spectra for Seismic Design of Nuclear Power Plants," dated December 1973

- RG 1.102, "Flood Protection for Nuclear Power Plants," dated September 1976

- RG 1.125, "Physical Models for Design and Operation of Hydraulic Structures and Systems for Nuclear Power Plants," dated March 2009

- RG 1.208, "A Performance-Based Approach to Define the Site-Specific Earthquake Ground Motion," dated March 2007

- RG 1.221, "Design-Basis Hurricane and Hurricane Missiles for Nuclear Power Plants," dated October 2011

The NRC staff continually evaluates new information on natural phenomena, including lessons learned from operational experience and their potential impact on risk and overall plant safety. These evaluations have led to new requirements or guidance as discussed above, updated regulatory guidance, generic communications, and plant-specific actions to address identified issues.

At the time of the Fukushima accident, the NRC staff was proceeding with regulatory actions to request licensees to evaluate updated seismic hazard information. In support of early site permits (ESP) and combined license (COL) applications for new reactors, the NRC staff reviewed updates to the seismic source and ground motion models the applicants provided. These reviews of the applications performed by the staff identified higher seismic hazard estimates than previously assumed, increasing the likelihood of exceeding the safe-shutdown earthquake at operating facilities in the Central and Eastern United States (CEUS). In 2005, the staff recommended an examination of increased seismic hazard estimates in the CEUS under the NRC Generic Issues Program (GIP). The NRC established Generic Issue (GI)-199, "Implications of Updated Probabilistic Seismic Hazard Estimates in Central and Eastern United States on Existing Plants," on June 9, 2005.

In 2010, the NRC concluded that GI-199 should transition to the regulatory assessment stage of the GIP. Information Notice (IN) 2010-018, "Generic Issue 199, Implications of Updated Probabilistic Seismic Hazard Estimates in Central and Eastern United States on Existing Plants," dated September 2, 2010, summarizes the results of the GI-199 safety and risk assessment. After it issued the IN, the NRC asked licensees to evaluate the updated seismic hazard analysis. The staff recommended, and the Commission approved, the incorporation of GI-199 into the regulatory actions being taken within the context of the JLD.

As a result of the NTTF review of the Fukushima events, the NRC concluded that seismic and flooding hazards warranted further consideration because of significant advancements in the state of knowledge and state of analysis in these areas in the time period since the operating plants were sited and licensed. One example of advancement in the state of knowledge is NUREG-2115, "Central and Eastern United States Seismic Source Characterization for Nuclear Facilities," dated January 2012, which presents updated composite seismic hazard curves for the CEUS that resulted from a joint NRC, Department of Energy (DOE) and Electric Power Research Institute (EPRI) effort. The Japan earthquake and subsequent tsunami also highlighted the need to evaluate concurrent related events, such as seismically-induced fires and floods. The NTTF identified a number of regulatory actions in the area of external events. These actions, as expanded or modified by the Steering Committee, are discussed below.

1.2 Seismic, Flooding, and Other Hazards Protection

1.2.1 Seismic and Flooding Hazards

1.2.1.1 *Discussion*

The NRC is undertaking near-term regulatory activities to reevaluate and upgrade, as necessary, the protection of SSCs against design-basis seismic and flooding events for all operating reactors in the United States (U.S.). These activities are based on NTTF Recommendations 2.1 and 2.3, as modified by subsequent NRC management direction. These activities include the following:

1. Request that licensees reevaluate the seismic and flooding hazards at their sites against current NRC requirements and guidance and, if necessary, identify actions to be taken to update the design basis and SSCs important to safety. (NTTF Recommendation 2.1)

2. Request that licensees perform seismic and flooding protection walkdowns to identify and address plant-specific vulnerabilities and verify the adequacy of monitoring and maintenance for protection features in the interim period until longer-term actions are completed to update the design basis for external events. (NTTF Recommendation 2.3)

1.2.1.2 *Actions Taken*

1. On March 23, 2011, the NRC issued TI 2515/183, "Followup to the Fukushima Daiichi Nuclear Station Fuel Damage Event," to its inspectors. The report focused on the adequacy of facility equipment, capabilities, and strategies for responding to large area fires, explosions, SBO events, and flooding. As such, the results of the inspections conducted by the NRC provided information regarding the readiness of licensees to respond to seismic and flooding events.

2. On May 11, 2011, the NRC issued Bulletin 2011-01, "Mitigating Strategies," to require that licensees provide a comprehensive verification of their compliance with the regulatory requirements for strategies required to maintain or restore core cooling, containment and spent fuel cooling capabilities under the circumstances associated with loss of large areas of the plant caused by explosions or fires. These strategies could be important in responding to a severe seismic or flooding event.

3. The NRC staff engaged stakeholders to discuss the technical basis and acceptance criteria for conducting a reevaluation of site-specific seismic hazards. In those

discussions, the staff considered the implementation of the hazard and risk methodologies described in draft GL 2011-XX, "Seismic Risk Evaluations for Operating Reactors," which was published in the *Federal Register* (FR Doc. 2011-22422, 76 FR 54507) on September 1, 2011.

After meetings with stakeholders, the NRC decided not to issue the GL in final form. Instead the staff decided that a more appropriate regulatory mechanism was to issue a request for information (RFI) pursuant to 10 CFR 50.54(f). The issuance of this RFI is discussed in Section 1.2.1.3.

4. On October 3, 2011, the NRC staff proposed that the NRC prioritize the NTTF recommendations to be taken in response to the lessons learned from Fukushima (SECY-11-0137, "Prioritization of Recommended Actions To Be Taken In Response to Fukushima Lessons Learned"). On December 15, 2011, the Commission approved the staff's proposed prioritization of the NTTF recommendations and stated its support for staff action on the near-term recommendations (staff requirements memorandum (SRM) on SECY-11-0137).

5. The NRC staff engaged stakeholders to inform NRC's process for defining guidelines for applying present-day regulatory guidance and methodologies being used for ESP and COL reviews to the reevaluation of seismic and flooding hazards at operating reactors. The NRC held the first public meeting on December 14, 2011, and the second one in January 18, 2012. The participants discussed RFIs associated with NTTF Recommendations 2.1 and 2.3, Flooding and Seismic Protections, including walkdown methodologies and acceptance criteria.

6. On January 13, 2012, the NRC Steering Committee again met with representatives from the Nuclear Energy Institute (NEI) and the nuclear industry to discuss how to implement near-term recommendations.

7. On March 12, 2012, the NRC issued RFIs asking licensees to (1) reevaluate site-specific seismic and flooding hazards using the updated methodology, and (2) identify actions they have taken -- or planned to take -- to address plant-specific issues associated with the updated seismic and flooding hazards (including potential changes to the licensing or design basis of a plant). (NTTF Recommendation 2.1) The specifics associated with the RFIs are discussed in SECY-12-0025, "Proposed Orders and Requests for Information in Response to Lessons Learned From Japan's March 11, 2011, Great Tohoku Earthquake and Tsunami," dated February 17, 2012.

8. On March 12, 2012, the NRC issued RFIs asking licensees to (1) identify and address plant-specific issues (through their corrective action program) and verify the adequacy of monitoring and maintenance for protection features by performing seismic and flooding walkdowns and (2) inform the NRC of the results of the walkdowns and corrective actions taken or planned. (NTTF Recommendation 2.3) This is discussed in SECY-12-0025.

9. In March 2012, the NRC staff completed its review of licensee responses to NRC Bulletin 2011-01.

1.2.1.3 *Actions Planned*

1. The NRC staff is in the process of determining whether regulatory actions are needed in light of the licensees' responses to NRC Bulletin 2011-01.

2. The NRC staff will prepare implementation guidance for seismic and flooding hazard walkdowns and guidance for seismic and flooding hazard reevaluations.

3. The NRS staff will evaluate each licensee's response to the RFIs and take appropriate regulatory action to resolve issues associated with updated site-specific seismic and flooding hazards.

4. The NRC will issue orders to licensees to require regulatory action, if necessary.

5. The NRC staff will conduct inspection activities and issue letters to close out the actions on a plant-by-plant basis.

1.2.1.4 *Schedule*

The staff's analysis of the responses to NRC Bulletin 2011-01 was completed in March 2012. Based on the analysis, the NRC will determine if further regulatory actions are needed by the summer 2012. The NRC staff will prepare implementation guidance for seismic and flooding hazard walkdowns by the end of May 2012. Guidance for seismic and flooding hazard reevaluations will be prepared by November 2012. The staff has not yet determined a schedule to evaluate the licensee responses to the RFIs on seismic and flooding reanalysis and walkdowns.

1.2.1.5 *Results*

None of the observations that resulted from the inspections conducted under TI 2515/183 indicated a significant safety issue; however, in general, they indicated a potential industry trend of failure to maintain the equipment and strategies needed to mitigate some beyond-design and design-basis events. A summary of the observations and the results overview are available at http://www.nrc.gov/NRR/OVERSIGHT/ASSESS/follow-up-rpts.html.

By June 11, 2011, all licensees had confirmed compliance with the regulatory requirements in Bulletin 2011-01. All licensees provided the requested information on maintenance, testing, offsite support and other features of their mitigating strategies programs. The staff has analyzed the licensees' responses and is in the process of determining whether additional regulatory actions are needed. Each licensee's response to this bulletin is available at http://www.nrc.gov/NRR/OVERSIGHT/ASSESS/mitigating-strategies.html.

The remainder activities are currently ongoing and results are not yet available.

1.2.2 Other External Hazards

1.2.2.1 *Discussion*

While undertaking actions to address the seismic and flooding hazards discussed above, the NRC staff recognized that it should reevaluate other external hazards against existing requirements and regulatory guidance. Other external hazards include phenomena such as tornados, hurricanes, severe winds, extreme temperatures, extreme precipitation, dust storms, forest fires, and volcanic activity. Thus, NTTF Recommendation 2.1 will be expanded to include consideration of other external hazards. The other external hazards evaluation will require significant resources for both licensees and NRC, as well as specialized expertise to review licensee re-evaluations and to document results of NRC evaluations. Since sufficient resource flexibility, including availability of critical skill sets, does not exist, the staff prioritized the other external hazards evaluation of NTTF Recommendation 2.1 as a Tier 2 activity.

1.2.2.2 *Actions Taken*

1. On October 3, 2011, the NRC staff proposed that the NRC prioritize the NTTF recommendations to be taken in response to the lessons learned from Fukushima (SECY-11-0137). On December 15, 2011, the Commission approved the staff's proposed prioritization of the NTTF recommendations (SRM on SECY-11-0137).

2. The NRC staff engaged stakeholders to inform the NRC's process for defining guidelines for applying present-day regulatory guidance and methodologies being used for ESP and COL reviews to the reevaluation of seismic and flooding hazards at operating reactors. The NRC held the first public meeting on December 14, 2011, and the second one on January 18, 2012.

3. On January 13, 2012, the NRC Steering Committee again met with representatives from NEI and the nuclear industry to discuss how to implement near-term recommendations, including reevaluation of other external hazards.

1.2.2.3 *Actions Planned*

Once sufficient expertise and resources are available, the staff will undertake regulatory activities to:

1. Continue stakeholder interactions to discuss the technical basis and acceptance criteria for conducting a re-evaluation of site-specific external natural hazards. These interactions will also help to define guidelines for the application of present-day regulatory guidance and methodologies being used for ESP and COL reviews to the re-evaluation of hazards at operating reactors.

2. Develop and issue a RFI to (1) re-evaluate other site-specific natural hazards using the methodology discussed in item 1 above, and (2) identify actions that have been taken, or are planned, to address plant-specific issues associated with the updated natural hazards (including potential changes to the licensing or design basis of a plant).

3. Evaluate licensee responses and take appropriate regulatory action to resolve issues associated with updated site-specific natural hazards.

1.2.2.4 *Schedule*

The NRC staff plans to develop and issue the RFIs on the reanalysis for other external hazards six months after sufficient expertise and resources are available. The staff has not yet determined a schedule to evaluate the licensee responses to the RFIs. Based on the results of these evaluations, the NRC will decide whether orders should be issued. If applicable, the NRC will conduct its inspection activities subsequent to the issuance and implementation of these orders.

1.2.2.5 *Results*

The activities are currently ongoing and results are not yet available.

1.3 State-of-the-Art Analysis

1.3.1 Discussion

Driven by technical advances since the last NRC-sponsored Level 3 probabilistic risk assessments (PRAs) were performed, the NRC staff is developing plans for a new Level 3 PRA. The last Level 3 PRA, NUREG-1150, "Severe Accident Risks: An Assessment for Five U.S. Nuclear Power Plants," was issued more than 20 years ago, in December 1990. Technical advances since that time include (1) plant modifications to enhance NPP operational performance, safety, and security; (2) improved understanding and modeling of severe accident phenomena; and (3) advances in PRA technology, such as common-cause modeling.

The staff also has identified additional scope considerations that could be addressed in a new and more comprehensive Level 3 PRA. These factors include (1) multi-unit site effects; (2) other site radiological sources (e.g., SFPs or dry storage casks); and (3) site-specific external hazards such as fires, flooding, and seismic events.

A new full-scope comprehensive site Level 3 PRA that incorporates these technical advances and additional scope considerations could improve the NRC's understanding of probable risk, enhancing regulatory decisionmaking, and helping the agency focus its limited resources on issues most pertinent to its mission to protect public health and safety. Two documents give a full description of Level 3 PRA activities: SECY-11-0089, "Options for Proceeding with Future Level 3 Probabilistic Risk Assessment Activities," dated July 7, 2011, and the associated SRM, dated September 21, 2011. In SECY-11-0089, the staff presented three options to the Commission for performing a Level 3 PRA: (1) maintain status quo and continue evolutionary development of PRA technology, (2) conduct focused research to address identified gaps in existing PRA technology before performing a full-scope comprehensive site Level 3 PRA, and (3) conduct a full-scope comprehensive site Level 3 PRA. In the SRM response, the Commission approved a modified version of Option 3 to conduct a full-scope comprehensive site Level 3 PRA for an operating plant. The modification to Option 3 extended the schedule to 4 years to alleviate some of the near-term resource challenges and allow adequate time for a careful site selection process. The new Level 3 PRA will offer insight into many of the NTTF recommendations.

1.3.2 Actions Taken

1. NRC is currently developing plans to perform a new multi-year Level 3 PRA to gain insights into multi-unit risk and total site risk. This assessment will be conducted for one multi-unit site.

1.3.3 Actions Planned

1. NRC will implement the plan and perform the Level 3 PRA.

1.3.4 Schedule

NRC will implement the plan and perform the Level 3 PRA within the 4-year time frame given by the Commission.

1.3.5 Results

The activities are currently ongoing and results are not yet available.

1.4 Events Beyond the Current Design Basis

1.4.1 Discussion

The staff's review of the NTTF recommendations identified areas for further evaluation to enhance the regulations and cope with events beyond the current design basis. Essentially all of the actions the NRC is pursuing relate to events beyond the current design basis. (Please see other sections of this report for discussions of station blackout (SBO), mitigating beyond-design-basis and multi-unit events, emergency preparedness (EP), containment venting, and SFP instrumentation.)

1.4.2 Actions Taken

1. On October 3, 2011, the NRC staff proposed that the NRC prioritize the NTTF recommendations to be taken in response to the lessons learned from Fukushima (SECY-11-0137). On December 15, 2011, the Commission approved the staff's proposed prioritization of the NTTF recommendations (SRM on SECY-11-0137).

2. The NRC currently is evaluating additional topics related to external events beyond the design basis. These topics include the following:

   - Rulemaking to require licensees to confirm seismic and flooding hazards every 10 years and address any new and significant information. (NTTF Recommendation 2.2)

   - Potential enhancements to licensees' capability to prevent or mitigate seismically-induced fires and floods. (NTTF Recommendation 3)

1.4.3 Actions Planned

1. The staff will write a SECY paper describing the plans and schedules for performing work on these additional topics.

1.4.4 Schedule

The staff is scheduled to deliver a SECY paper describing the plans and schedules for performing work on these additional topics to the Commission in July 2012.

1.4.5 Results

The activities are currently ongoing and results are not yet available.

TOPIC 2. DESIGN ISSUES

2.1 Introduction

After the March 2011 accident at the Fukushima Daiichi NPP, the NRC decided to take specific regulatory actions in areas of NPP design to improve the availability and reliability of plant safety systems. These actions include additional protection against losses of one or more of the following:

- all onsite and offsite alternating current (AC) power sources

- containment heat removal and overpressure protection

- SFP cooling

- ultimate heat sink (UHS)

These actions will build upon existing NRC regulations and will help ensure the continued availability and reliability of NPP systems. Topic 2 of this report describes specific actions the NRC will take.

The availability of AC power is essential for the safe operation and accident recovery in all NPPs currently in use in the U.S. All NPPs are connected to the electrical grid which supplies its offsite AC power source. If a plant experiences a loss of offsite power (LOOP), emergency diesel generators (EDGs) provide onsite AC power. If the EDGs also fail, resulting in a total loss of AC power, the plant will experience an SBO. Certain passive, new reactor designs, such as the Advanced Passive 1000 (AP1000) and Economic Simplified Boiling Water Reactor (ESBWR), can cope for a significant period of time (72 hours) without AC power.

The unavailability of AC power can have a significant adverse impact on an NPP's ability to achieve and maintain safe shutdown conditions. Risk analyses performed for U.S. NPPs indicate that the loss of all AC power can be a significant contributor to the risk associated with plant operation, contributing more than 70 percent of the overall risk at some plants. Therefore, a LOOP and its subsequent restoration are important inputs to plant risk models. Other important contributors to the plant's risk during LOOP-initiated scenarios involve situations in which plants must achieve safe shutdown by relying on components that do not require AC power, such as turbine- or diesel-driven pumps. Thus, the reliability of such components, the capacity of direct current (DC) batteries, and the timeliness of offsite power restoration are important contributors to SBO risk.

In 1980, the NRC established a plan to address concerns about SBO risk and the associated reliability of EDGs. In August 1988, the NRC issued the SBO rule (10 CFR 50.63, "Loss of All Alternating Current Power") and the associated RG 1.155, "Station Blackout." The SBO rule requires that NPPs have the capability to withstand a SBO and maintain core cooling for a specified duration. The method described in RG 1.155 results in a minimum acceptable SBO duration coping capability ranging from 2 to 16 hours. The result for all U.S. operating plants was a coping duration of either 4 or 8 hours. NPPs also were required to enhance procedures and training for restoring both offsite and onsite AC power sources. To meet the requirements of the SBO rule, some licensees chose to make NPP modifications, such as adding more sources of emergency AC power.

To address protection from loss of containment cooling and containment overpressure, the NRC issued a generic communication to boiling water reactor (BWR) licensees on the safety benefits of installing hardened wetwell vents (GL 89-16, "Installation of a Hardened Wetwell Vent," dated September 1, 1989). All BWR licensees with Mark I containments voluntarily installed a hardened vent. Also, to date, three of eight BWR Mark II units have installed hardened vents. No regulatory requirement was imposed at that time, and designs of these systems varied – some relied on AC-operated, DC-operated, or air-operated valves; some incorporated passive rupture disks along with isolation valves.

The NRC has taken additional steps over the years to require reliable equipment and strategies to provide adequate cooling to reactors and SFPs. The NRC initiated these requirements after the terrorist attacks of September 11, 2001. These requirements are now codified in 10 CFR 50.54(hh)(2), and both the NRC and the U.S. nuclear industry have developed guidance to implement this rule. To comply with the requirements, each NPP licensee expanded its command and control capabilities and developed strategies and capabilities to cool fuel in the reactor and SFP and to mitigate releases without having to rely on the site's AC electrical power distribution system.

As a result of the Japanese tsunami, the reactors at Fukushima Daiichi lost the capacity to release the heat being produced by the cores of units 1, 2 and 3 to the UHS (the ocean). The event reinforces the need to evaluate the capacity to restore an UHS promptly under accident conditions, and to include in accident planning consideration an alternative means for maintaining the reactor stable in hot standby for an extended period of time when normal modes of heat transport to the UHS are unavailable. The NRC has taken regulatory actions as a result of the lessons learned from Fukushima that addresses several aspects of the loss of UHS.

The NRC's review of the Fukushima Daiichi accident identified a number of NPP design issues. The activities discussed below focus on regulatory actions that the NRC has taken (or plans to take) to enhance a plant's capability to prevent significant core damage if it experiences a beyond-design-basis external hazard.

2.2 Preventing Loss of Onsite and Offsite Alternating Current Power Sources (Station Blackout)

2.2.1 Discussion

The NRC issued orders to licensees to develop, implement, and maintain guidance and strategies to mitigate the effects of challenges to the key safety functions of core cooling, containment, and SFP cooling capabilities for beyond-design-basis events and multi-unit events. (NTTF Recommendation 4.2 as expanded by the Steering Committee). In addition, the NRC is planning regulatory activities that will address design enhancements to further reduce the probability of SBO, including a rulemaking to improve the SBO rule (10 CFR 50.63) (NTTF Recommendation 4.1).

2.2.2 Actions Taken

1. On March 23, 2011, the NRC issued TI 2515/183 to provide guidance to its inspectors to assess the adequacy of actions taken by U.S. licensees in response to the Fukushima

14

accident. The intent was to provide the NRC a high-level look at the U.S. industry's preparedness for events that may exceed the design basis for a plant.

2. On October 3, 2011, the NRC staff proposed that the NRC prioritize the NTTF recommendations to be taken in response to the lessons learned from Fukushima (SECY-11-0137). On December 15, 2011, the Commission approved the staff's proposed prioritization of the NTTF recommendations (SRM on SECY-11-0137).

3. On March 12, 2012, the NRC issued orders to provide for reasonable protection of the equipment needed to maintain or restore core cooling, containment, and SFP cooling capabilities in the event that explosions or fires cause the loss of large areas of the plant. These orders are discussed in SECY-12-0025.

4. On March 20, 2012, the NRC issued an advance notice of proposed rulemaking (ANPR) to engage stakeholders in rulemaking activities associated with the potential SBO rulemaking (NTTF Recommendation 4.1). The ANPR will allow the NRC to obtain stakeholder input early in the rulemaking process.

2.2.3 Actions Planned

1. Undertake a rulemaking to improve the SBO rule. In the rulemaking process, the NRC will develop a regulatory basis, issue a proposed rule and supporting guidance, resolve public comments, and issue the final rule and guidance documents. A period for licensees to implement the rule and submit license amendments will follow. The NRC will inspect sites to ensure that licensees meet the new regulations. (NTTF Recommendation 4.1)

2. The NRC staff is planning to engage stakeholders to (1) inform the development of acceptance criteria for reasonable protection of equipment needed to respond to beyond-design-basis external hazards and multi-unit events, (2) assess the need to supplement equipment to support beyond-design-basis and multi-unit event mitigation, and (3) discuss the need to develop and provide training on supporting strategies.

3. The NRC staff intends to develop plant-specific safety evaluation reports and perform inspections to verify compliance with the orders.

2.2.4 Schedule

The SBO rulemaking is expected to be completed within 2.5 years (approximately December 2014). The full implementation of orders -- to provide for reasonable protection of the equipment needed to maintain or restore core cooling, containment, and SFP cooling capabilities under the circumstances associated with beyond-design-basis events and multi-unit events -- is expected to be complete no later than two refueling cycles after February 28, 2013 (the date by which licensees must submit their plans to address the orders), or by December 31, 2016, whichever comes first. The NRC staff has not yet developed schedules for its inspection activities.

2.2.5 Results

None of the observations that resulted from the inspections conducted under TI 2515/183 indicated a significant safety issue; however, in general, they indicated a potential industry trend of failure to maintain the equipment and strategies needed to mitigate some beyond-design and design-basis events. A summary of the observations and the results overview are available at http://www.nrc.gov/NRR/OVERSIGHT/ASSESS/follow-up-rpts.html.

2.3 Containment Overpressure Protection

2.3.1 Discussion

The NRC issued orders to all licensees of NPPs with Mark I or Mark II containment designs to provide reliable hardened wetwell vents. In addition, the NRC is planning three long-term regulatory activities to ensure protection of NPP containment functions. These activities include the following:

1. Consider regulatory action for hardened vents for other containment designs (NTTF Recommendation 5.2).

2. Evaluate enhanced measures for hydrogen control and mitigation (NTTF Recommendation 6).

3. Consider adding filtered vents for Mark I and Mark II containments (NTTF Recommendation 5.1), such that further analysis and interaction with stakeholders will inform whether filtered vents should be required. The staff has determined that consideration of severe accident conditions in the design and operation of the vents and the possible addition of filters to the vents are significant policy issues that will require further consideration and stakeholder interaction for the Commission to make an informed decision.

2.3.2 Actions Taken

1. On October 3, 2011, the NRC staff proposed that the NRC prioritize the NTTF recommendations to be taken in response to the lessons learned from Fukushima (SECY-11-0137). On December 15, 2011, the Commission approved the staff's proposed prioritization of the NTTF recommendations (SRM on SECY-11-0137).

2. The NRC engaged stakeholders to help the NRC determine the technical bases and acceptance criteria for suitable design expectations of hardened wetwell vents.

3. On March 12, 2012, the NRC issued orders to all licensees of NPPs with Mark I or Mark II containment designs to provide reliable hardened wetwell vents. These orders are discussed in SECY-12-0025. (NTTF Recommendation 5.1).

2.3.3 Actions Planned

1. The NRC staff will perform inspections to verify compliance with the orders.

2. NRC is planning to engage with stakeholders to discuss reliable hardened vents and determine the necessity for taking regulatory actions to require filtration of containment vents and provide the Commission with a staff recommendation.

3. The NRC is planning to conduct a longer-term evaluation of the need for reliable hardened vents in containment designs that are not Mark I and Mark II. (NTTF Recommendation 5.2)

4. The NRC is planning to conduct a longer-term evaluation of the need for enhanced hydrogen control and mitigation inside containment or other buildings. (NTTF Recommendation 6)

2.3.4 Schedule

The orders regarding the implementation of reliable hardened wetwell vents for Mark I and Mark II containments should be fully implemented no later than two refueling cycles after February 28, 2013 (the date by which licensees must submit their plans to address the orders), or by December 31, 2016, whichever comes first. The NRC staff has not yet developed schedules for its inspection activities.

The NRC staff expects to issue a SECY paper with its recommendation on whether filtration of containment vents should be required by July 2012.

The NRC staff has not yet developed schedules for longer-term activities to consider hardened vents for other containment designs (NTTF Recommendation 5.2) or to evaluate enhanced measures to control and mitigate hydrogen (NTTF Recommendation 6).

2.3.5 Results

The activities are currently ongoing and results are not yet available.

2.4 Reliable Cooling of Spent Fuel Pools

2.4.1 Discussion

The NRC issued an order to all operating NPP licensees to provide reliable instrumentation to measure SFP water levels. In addition, the NRC is planning a long-term regulatory action to ensure continued cooling of spent fuel in the SFPs. This activity includes the issuance of a rulemaking to provide reliable instrumentation and makeup capabilities (NTTF Recommendations 7.2 through 7.5).

2.4.2 Actions Taken

1. On October 3, 2011, the NRC staff proposed that the NRC prioritize the NTTF recommendations to be taken in response to the lessons learned from Fukushima (SECY-11-0137). On December 15, 2011, the Commission approved the staff's proposed prioritization of the NTTF recommendations (SRM on SECY-11-0137).

2. The NRC engaged stakeholders to help the NRC staff determine (1) what constitutes reliable (potentially safety-related) SFP instrumentation, (2) what conditions the instrumentation must withstand to fulfill its intended function, and (3) where indications are needed (e.g., control room and/or remote location).

3. On March 12, 2012, the NRC issued orders to all operating NPP licensees to provide reliable indicators of water level in the SFP. These orders are discussed in SECY-12-0025.

2.4.3 Actions Planned

1. The NRC staff will develop plant specific safety evaluation reports and perform inspections to verify compliance with the orders.

2. Undertake a rulemaking to provide for enhanced SFP makeup capabilities (NTTF Recommendations 7.2 through 7.5), and reliable SFP water level instrumentation (derived NTTF Recommendation 7.1).

2.4.4 Schedule

The NRC expects that full implementation of the orders to provide reliable SFP instrumentation should be completed no later than two refueling cycles after February 28, 2013 (the date by which licensees must submit their plans to address the orders), or by December 31, 2016, whichever comes first. Subsequent to the implementation of the orders, the NRC will conduct its inspection activities. The NRC staff has not yet developed the inspection schedule. The NRC expects to complete the rulemaking related to SFP makeup capabilities and reliable water level instrumentation within 5 years.

2.4.5 Results

The activities are currently ongoing and results are not yet available.

2.5 Preventing Loss of Ultimate Heat Sink

2.5.1 Discussion

The NRC is planning two short-term regulatory activities to further reduce the risk of plants losing access to their normal UHS. Preventing the loss of UHS was not included in the recommendations from the NTTF report, but the NRC later identified this issue as one needed to be considered and included it for consideration in SECY-11-0137.

2.5.2 Actions Taken

1. On March 23, 2011, the NRC issued TI 2515/183, "Followup to the Fukushima Daiichi Nuclear Station Fuel Damage Event," to its inspectors. The report focused on the adequacy of facility equipment, capabilities, and strategies for responding to large area fires, explosions, SBO events, and flooding.

2. On October 3, 2011, the NRC staff proposed that the NRC prioritize the NTTF recommendations to be taken in response to the lessons learned from Fukushima (SECY-11-0137). On December 15, 2011, the Commission approved the staff's proposed prioritization of the NTTF recommendations (SRM on SECY-11-0137).

3. The NRC held a public meeting on December 1, 2011, with members of the public, NEI, and industry, to discuss the implementation of the NTTF recommendations. During this

meeting, the NRC communicated that design basis natural event could result in interruption of access to the normal UHS and that this issue would be considered during the resolution of near-term regulatory actions.

4. On March 12, 2012, the NRC issued orders to all licensees to provide mitigating measures for beyond-design-basis external events that result in a loss of UHS in conjunction with the resolution of NTTF Recommendation 4.2. This NRC order is discussed in SECY-12-0025.

2.5.3 Actions Planned

1. The NRC plans to include UHS as an important system to consider during the resolution of the near-term regulatory actions as described below:

    a. The NRC plans to request licensees to include UHS in the reevaluation and walkdowns of site-specific seismic and flooding hazards, and identify actions taken or planned to address plant-specific issues associated with the updated seismic hazards in conjunction with the resolution of NTTF Recommendations 2.1 and 2.3. These RFIs are discussed in SECY-12-0025.

    b. The NRC plans to incorporate the loss of UHS as a design assumption during the SBO rulemaking activities in conjunction with the resolution of NTTF Recommendation 4.1.

    c. The NRC plans to address loss of UHS in the context of the reevaluation of other external hazards as described in Section 1.2.2.3.

2.5.4 Schedule

The NRC staff has not yet developed the schedule for evaluating the RFI responses. Based on the results of these evaluations, the NRC will decide whether orders should be issued.

The SBO rulemaking, which will incorporate loss of UHS, is expected to be completed within 2.5 years.

The implementation of orders -- to provide equipment needed to maintain or restore core cooling, containment, and SFP cooling capabilities under the circumstances associated with beyond-design-basis events and multi-unit events -- is expected to be complete no later than two refueling cycles after February 28, 2013 (the date by which licensees must submit their plans to address the orders), or by December 31, 2016, whichever comes first. The NRC staff has not yet developed the schedule for its inspection activities.

2.5.5 Results

None of the observations that resulted from the inspections conducted under TI 2515/183 indicated a significant safety issue; however, in general, they indicated a potential industry trend of failure to maintain the equipment and strategies needed to mitigate some beyond-design and design-basis events. A summary of the observations and the results overview are available at http://www.nrc.gov/NRR/OVERSIGHT/ASSESS/follow-up-rpts.html.

The remainder activities are currently ongoing and results are not yet available.

TOPIC 3. SEVERE ACCIDENT MANAGEMENT AND RECOVERY (ONSITE)

3.1 Introduction

After the March 2011 accident at the Fukushima Daiichi NPP, the NRC decided to take specific regulatory actions in areas of severe accident management and onsite recovery.   At U.S. NPPs, combinations of onsite programs contribute to severe accident management and overall emergency response capability.   These programs include abnormal operating procedures (AOPs), alarm response procedures, emergency operating procedures (EOPs), severe accident management guidelines (SAMGs), and extensive damage mitigation guidelines (EDMGs).   The technical basis and rationale for these procedures represent lessons learned from international experiences; U.S. reactor accidents; and the September 11, 2001, terrorist attacks.   What follows is a brief history of the NRC's current regulatory approach to severe accident management and the NRC's plans to develop additional requirements to address lessons learned from the Fukushima accident.

In October 1980, following the March 1979 accident at Three Mile Island (TMI) Unit 2, the NRC staff developed guidance for NPP licensees to develop comprehensive and integrated plans to improve safety at power reactors (NUREG-0660, "NRC Action Plan Developed as a Result of the TMI-2 Accident", dated May 1980).   The Commission approved specific items from this guidance for implementation at nuclear power reactors (NUREG-0737, "Clarification of TMI Action Plan Requirements," dated November 1980).   In January 1983, the NRC published additional supplemental guidance, which included its expectations for licensees to upgrade their EOPs (NUREG-0737, Supplement 1, "Clarification of TMI Action Plan Requirements: Requirements for Emergency Response Capability," dated January 1983).   EOPs are used by reactor operators to mitigate the consequences of transients and accidents that cause plant parameters to exceed setpoints for the reactor protection system or the actuation of engineered safety features.   In response to the 1983 guidance, the NRC required licensees to reanalyze transients and accidents and prepare technical guidelines. These analyses identify operator tasks as well as information and control needs.   The analyses also serve as the basis for integrating upgraded EOPs and the control room design review and for verifying the design of safety parameter display systems.

The nuclear industry also voluntarily developed SAMGs in response to the TMI accident; details can be found in GL 88-20, Supplement 2, "Accident Management Strategies for Consideration in the Individual Plant Examination Process," dated April 4, 1990. However, the NRC did not require licensees to develop and implement SAMGs. The SAMGs are used by plant technical support staff in the plant's technical support center, when an accident at the plant progresses beyond the point at which EOPs and other plant procedures are applicable. The SAMGs are voluntary, and training and licensing of plant operators may not address them.

After the terrorist attacks of September 11, 2001, the NRC promptly issued advisories that licensees should demonstrate their ability to deal with losses of large areas of their plants caused by fires and explosions. These early security advisories and orders are now contained in a new regulation at 10 CFR 50.54(hh). This rule includes requirements for EDMGs. Like SAMGs, the EDMGs would be used in the unlikely event that the normal command and control structure is disabled and the continued use of EOPs is no longer feasible.

Following the March 2011 accident at Fukushima Daiichi, the NRC recognized that each of the onsite emergency action programs described above contributes to the overall capability of plant operators to mitigate accidents in response to an emergency. The SAMGs and EDMGs complement the EOPs in an important way and have established the command and control responsibilities for each of these programs, although not necessarily in a consistent manner. As described above, each of these programs was developed at a different time to serve a different purpose, and each of these programs is treated differently in the NRC's regulations, inspection program, and licensing process, as well as in licensee programs and organizations.

The NRC has identified a number of regulatory actions in the area of severe accident management and recovery (onsite) as a result of its review of the Fukushima Daiichi accident. The specific actions are discussed below.

3.2 U.S. Nuclear Regulatory Commission Personnel Resources and Training

3.2.1 Discussion

The NRC estimates that issuing new regulations over the next 5 years will require a diverse and highly specialized professional staff at the NRC with experience in such matters as reactor licensing, seismology, structural engineering, PRA, geotechnical engineering, inspection program management, inspection, surface water hydrology, mechanical engineering, instrumentation and control, electrical engineering, fire protection, nuclear engineering, incident response and EP, and rulemaking.

The NRC's significant challenge will be to ensure that regulatory improvements in response to Fukushima do not displace ongoing work that has greater safety benefit. The NRC also acknowledges that there are resource and implementation challenges that licensees will experience with regard to skill sets in high demand (e.g., PRA, seismic, and flooding expertise).

3.2.2 Actions Taken

1. On October 3, 2011, the NRC staff proposed that the NRC prioritize the NTTF recommendations to be taken in response to the lessons learned from Fukushima (SECY-11-0137). On December 15, 2011, the Commission approved the staff's proposed prioritization of the NTTF recommendations (SRM on SECY-11-0137).

3.2.3 Actions Planned

1. The NRC has identified a longer-term activity to train NRC staff on severe accidents and train resident inspectors on SAMGs (NTTF Recommendation 12.2). Other specific regulatory actions are described below under Sections 3.3 and 3.5.

3.2.4 Schedule

The NRC plans to develop a schedule for NRC staff training on SAMGs.

3.2.5 Results

The activities are currently ongoing and results are not yet available.

3.3 Adequacy of Procedures

3.3.1 Discussion

The March 2011 accident at the Fukushima Daiichi site demonstrated the need for better integration of EOPs, SAMGs, and EDMGs. (NTTF Recommendation 8). In particular, these improved procedures should address the possibility of severe external hazards creating a prolonged SBO condition, which affects multiple reactor units and SFPs (NTTF Recommendation 9).

The NRC concluded that further integration could substantially enhance the overall effectiveness of those programs. The integration should clarify transition points, command and control, and decisionmaking, and rigorously train staff using conditions that are as close to real accident conditions as feasible (NTTF Recommendation 8.1). The NRC also concluded that action is warranted to confirm, augment, consolidate, simplify, and strengthen current regulatory and industry programs in a manner that produces a single, comprehensive framework for accident mitigation. Once the regulatory framework has been expanded, the NRC plans to modify its Reactor Oversight Process for inspection and performance monitoring of licensees to incorporate the expanded framework (NTTF Recommendation 12.1).

The framework may specify the authority to implement SAMGs and EDMGs without the need for a licensee to seek the NRC's permission or to invoke the regulations in 10 CFR 50.54(x) and (y), which allow operators to take action in an emergency that departs from a license condition or technical specification because it is necessary to protect public safety. This change would further clarify authority, streamline decisionmaking, and prevent potential delays in taking important emergency actions (included within NTTF Recommendation 8).

The NRC plans to issue an ANPR to engage stakeholders in rulemaking activities associated with the methods to integrate onsite emergency response processes, procedures, training and exercises. The ANPR will address a methodology and acceptance criteria to do the following (all related to NTTF Recommendations 8.1 through 8.4):

1. Modify the EOP technical guidelines (required by NUREG-0737, Supplement 1, "Requirements for Emergency Response Capability").

2. Include EOPs, SAMGs, and EDMGs in an integrated manner.

3. Specify clear command and control strategies for their implementation.

4. Stipulate appropriate qualification and training for those who make decisions during emergencies.

5. Modify the standard technical specifications for each operating reactor design to reference the approved EOP technical guidelines for that plant.

6. Require licensees to modify each plant's technical specifications to conform to these changes.

7. Require more realistic, hands-on training and exercises on SAMGs and EDMGs for all staff expected to implement the strategies and make decisions.

3.3.2 Actions Taken

1. On March 18, 2011, the NRC issued an IN to licensees and new reactor applicants. IN 2011-05, "Tohoku-Taiheiyou-Oki Earthquake Effects On Japanese Nuclear Power Plants," notified addressees of the effects of the Japanese Earthquake on NPPs and asked licensees to consider actions at their plants that would avoid similar problems. the NRC sent a similar notice to fuel cycle facilities on March 31, 2011 (IN 2011-08, "Tohoku-Taiheiyou-Oki Earthquake Effects on Japanese Nuclear Power Plants – For Fuel Cycle Facilities").

2. On April 29, 2011, the NRC issued TI 2515/184, "Availability and Readiness of SAMGs" to provide guidance to its inspectors on assessing the availability and readiness of SAMGs.

3. On May 11, 2011, the NRC issued Bulletin 2011-01, "Mitigating Strategies," to require that licensees provide a comprehensive verification of their compliance with the regulatory requirements for strategies required to maintain or restore core cooling, containment and spent fuel cooling capabilities under the circumstances associated with loss of large areas of the plant caused by explosions or fires.

4. On October 3, 2011, the NRC staff proposed that the NRC prioritize the NTTF recommendations to be taken in response to the lessons learned from Fukushima (SECY-11-0137). On December 15, 2011, the Commission approved the staff's proposed prioritization of the NTTF recommendations (SRM on SECY-11-0137).

5. In March 2012, the NRC staff completed the analysis of the licensee's responses to NRC Bulletin 2011-01.

6. On April 18, 2012, the NRC issued an ANPR to engage stakeholders in rulemaking activities associated with the methodology for integration of onsite emergency response process, procedures, training, and exercises (NTTF Recommendation 8, see Section 3.3.1). The ANPR will allow the NRC to obtain stakeholder input early in the rulemaking process.

3.3.3 Actions Planned

1. The NRC staff will use the analysis of the licensees' responses to NRC Bulletin 2011-01 to inform the development of guidance for the recently issued "Order Modifying Licenses with Regard to Requirements for Mitigation Strategies for Beyond-Design-Basis External Events."

2. In addition, the rulemaking process will develop a regulatory basis, issue a proposed rule and supporting guidance, resolve public comments, and issue the final rule and guidance documents. A period for licensees to implement the rule and submit license amendments will follow. The NRC will inspect sites to ensure licensees meet the new regulations.

3.3.4 Schedule

The staff's analysis of the responses to NRC Bulletin 2011-01 was completed in March 2012. Based on the analysis, the staff is informing the development of the guidance for the recently issued order and is in the process of determine whether further regulatory actions are needed. The staff's conclusion is expected by the summer 2012 and the guidance document by August 2012. The final rule is scheduled to be issued in approximately 5 years. The SECY paper describing options for expanding the regulatory framework to include beyond-design-basis requirements (NTTF Recommendation 1) is scheduled to be issued by February 2013. The NRC has yet to determine the schedule for making changes to the Reactor Oversight Process (NTTF Recommendation 12.1).

3.3.5 Results

None of the observations that resulted from the inspections conducted under TI 2515/184 indicated a significant safety issue; however, in general, they indicated an inconsistent implementation of some aspects of the voluntary SAMG program. A summary of the observations and the results overview are available at http://www.nrc.gov/NRR/OVERSIGHT/ASSESS/SAMGs.html.

By June 11, 2011, all licensees had confirmed compliance with the regulatory requirements in Bulletin 2011-01. All licensees had provided the requested information on maintenance, testing, offsite support, and other features of their mitigating strategies programs. The staff has analyzed the licensees' responses and is in the process of determining whether further regulatory actions are needed. Each licensee's response to this bulletin is available at http://www.nrc.gov/NRR/OVERSIGHT/ASSESS/mitigating-strategies.html.

3.4 Multi-Unit Events

3.4.1 Discussion

As described in Section 3.3, a methodology to better integrate the various types of procedures (i.e., EOPs, SAMGs, EDMGs) will be developed as part of the rulemaking and will apply to both single-unit and multi-unit sites. However, actions have been taken to require licensees to develop, implement, and maintain guidance and strategies to mitigate the effects of challenges to the key safety functions of core cooling, containment, and SFP cooling capabilities for beyond-design-basis events and multi-unit events. (NTTF Recommendation 4.2 as expanded by the Steering Committee). The NRC is also planning to develop a rulemaking to integrate EOPs, SAMGs, and EDMGs. (NTTF Recommendation 8). Further, the NRC is currently developing plans to perform a new multi-year Level 3 PRA to gain insights into multi-unit risk and total site risk.

3.4.2 Actions Taken

1. On December 15, 2011, the Commission approved a prioritization of recommended actions to be taken in response to the lessons learned from Fukushima (SRM on SECY-11-0137).

2. The NRC staff engaged stakeholders to (1) inform the development of acceptance criteria for reasonable protection of equipment needed to respond to beyond-design-basis external hazards and multi-unit events, (2) assess the need to supplement equipment to support beyond-design-basis and multi-unit event mitigation, and (3) discuss the need to develop and train licensee personnel on supporting strategies.

3. On March 12, 2012, the NRC issued orders to all operating NPP licensees requiring them to take steps to provide reasonable protection of equipment needed for strategies to mitigate beyond-design-basis events and multi-unit events. These orders are discussed in SECY-12-0025. (NTTF Recommendation 4.2).

3.4.3 Actions Planned

1. The NRC staff will develop plant specific safety evaluation reports and perform inspections to verify compliance with the orders.

3.4.4 Schedule

The full implementation of these orders is expected no later than two refueling cycles after February 28, 2013 (the date by which licensees must submit their plans to address the orders), or by December 31, 2016, whichever comes first. As stated in Section 3.3, the final rule pertaining to integration of onsite emergency response processes, procedures, training, and exercises is scheduled to be issued in approximately 5 years.

3.4.5 Results

The activities are currently ongoing and results are not yet available.

3.5 Equipment Availability

3.5.1 Discussion

Some of the near-term activities that the NRC is specifically undertaking to enhance onsite severe accident prevention are focused on improving equipment availability. These enhancements may also provide benefits with respect to severe accident mitigation and recovery actions at U.S. NPPs. These near-term activities are listed, along with other planned activities, in Section 3.5.3.

3.5.2 Actions Taken

1. On March 23, 2011, the NRC issued TI 2515/183, "Followup to the Fukushima Daiichi Nuclear Station Fuel Damage Event," to its inspectors. The report focused on the adequacy of facility equipment, capabilities, and strategies for responding to large area fires, explosions, SBO events, and flooding. Although the focus of these inspections was on equipment intended to prevent a severe accident, the equipment inspected would also provide benefits with respect to severe accident mitigation and recovery actions.

2. On October 3, 2011, the NRC staff proposed that the NRC prioritize the NTTF recommendations to be taken in response to the lessons learned from Fukushima (SECY-11-0137). On December 15, 2011, the Commission approved the staff's proposed prioritization of the NTTF recommendations (SRM on SECY-11-0137).

3. On March 12, 2012, the NRC issued orders to provide equipment required to maintain or restore core cooling, containment and spent fuel cooling capabilities under the circumstances associated with beyond-design-basis events and multi-unit events (NTTF

Recommendation 4.2). These orders are discussed in SECY-12-0025.

4. On March 12, 2012, the NRC issued an order to licensees that operate BWRs with Mark I and Mark II containment designs, to ensure reliable operation of hardened wetwell vents (NTTF Recommendation 5.1). This near-term order is discussed in SECY-12-0025.

5. On March 12, 2012, the NRC issued an order to licensees to provide reliable SFP instrumentation and procedures for operation, maintenance, and testing of the equipment (NTTF Recommendations 7.1). The near-term order is discussed in SECY-12-0025.

3.5.3 Actions Planned

1. A near-term rulemaking to enhance the capability of stations to maintain safety through a prolonged SBO (NTTF Recommendation 4.1).

2. A longer term activity to consider regulatory action for hardened vents for other containment designs (NTTF Recommendation 5.2).

3. A longer term activity to evaluate the need for enhanced measures for hydrogen control and mitigation (NTTF Recommendation 6).

3.5.4 Schedule

The full implementation of these near-term orders is expected no later than two refueling cycles after February 28, 2013 (the date by which licensees must submit their plans to address the orders), or by December 31, 2016, whichever comes first. All final rules are scheduled to be completed within 5 years. No schedules have been developed for the longer-term activities.

3.5.5 Results

None of the observations that resulted from the inspections conducted under TI 2515/183 indicated a significant safety issue; however, in general, they indicated a potential industry trend of failure to maintain the equipment and strategies needed to mitigate some beyond-design and design-basis events. A summary of the observations and the results overview are available at http://www.nrc.gov/NRR/OVERSIGHT/ASSESS/follow-up-rpts.html.

The remainder activities are currently ongoing and results are not yet available.

TOPIC 4. NATIONAL ORGANIZATIONS

4.1 Introduction

The U.S. Congress created the NRC as an independent regulatory agency in January 1975, with the passage of the Energy Reorganization Act. By giving the NRC an exclusively regulatory mandate, the statute reflected (in part) a congressional judgment that the expanding commercial nuclear power industry (which was expected to continue to grow) warranted the full-time attention of an exclusively regulatory agency. In creating the NRC, Congress also addressed a developing public concern that regulatory responsibilities were overshadowed by the promotion of nuclear power at the Atomic Energy Commission.

The NRC is headed by a five-member Commission. The President designates one member to serve as Chairman and official spokesperson. The Commission as a whole formulates policies and regulations governing nuclear reactor and materials safety, issues orders to licensees, and adjudicates legal matters brought before it. The Executive Director for Operations (EDO) carries out the policies and decisions of the Commission and directs the activities of the program offices. As a result of the Fukushima event, a new organization, the JLD, was formed to perform a longer-term review of the March 11, 2011, Japanese earthquake and tsunami. This directorate is part of the Office of Nuclear Reactor Regulation (NRR), which reports to the EDO. The JLD coordinates all of the NRC actions in response to Fukushima.

The NRC has frequent contact and interaction with the executive branch, including the White House, the Office of Management and Budget, U.S. Department of State (DOS), DOE, U.S. Environmental Protection Agency (EPA), Federal Emergency Management Agency (FEMA), U.S. Department of Labor, U.S. Department of Transportation, U.S. Department of Defense, U.S. Department of Homeland Security (DHS), National Oceanic and Atmospheric Administration (NOAA), and U.S. Department of Justice. After the Fukushima event, the NRC continued, and in some cases increased, communications with these organizations.

4.2 Regulator

The NRC's mission is to ensure the safe use of radioactive materials for beneficial civilian purposes while protecting people and the environment. The NRC was established by the Energy Reorganization Act of 1974, under which the agency took over regulatory responsibility for civilian nuclear facilities, formerly carried out by the Atomic Energy Commission (as established in the Atomic Energy Action of 1954). The NRC's regulations are contained in the Title 10 of the *Code of Federal Regulations*, "Energy." The NRC Commissioners are appointed by the President and confirmed by the Senate for 5-year terms. As a collegial body, the Commission formulates policies, develops regulations governing the safety of nuclear reactors and nuclear material, issues orders to licensees, and adjudicates legal matters. The staff of the NRC provides technical assessments and recommendations to aid the Commission's decisionmaking.

In response to the events at the Fukushima Daiichi NPP on March 11, 2011, the NRC established the NTTF. This group, consisting of senior managers and NRC officials, conducted a systematic and methodical review of NRC regulations to determine whether the agency should make additional improvements to its regulatory system in light of the events that unfolded. The NTTF identified 12 recommendations that are intended to clarify and strengthen the regulatory framework for protection against natural disasters. In some cases, subsequent recommendations from the Steering Committee have modified or enhanced the NTTF recommendations. The NRC then tasked its staff with prioritizing and implementing the NTTF recommendations. Each NTTF recommendation was prioritized by the Steering Committee into one of three groups depending on such factors as significance of risk and benefit to safety. The NRC also took certain immediate actions, in the form of an IN, TIs, a bulletin, and inspections, to ensure that no immediate safety concerns existed at American facilities. The JLD is responsible for overseeing the implementation of NTTF recommendations, identifying and evaluating additional recommendations, and overseeing the development of plans to address the items identified as requiring long-term review.

4.2.1 Information Notices and Temporary Instructions Issued in Response to the Effects of the Earthquake and Tsunami in Japan

4.2.1.1 Discussion

One week after the event, the NRC issued an IN to alert licensees of the effects of the Tohoku-Taiheiyou-Oki earthquake and resultant tsunami on NPPs in Japan. The NRC expected licensees to review the information for applicability to their facilities and consider actions, as appropriate.

In addition, the NRC gave its inspectors guidance on how to assess the following:

- Actions taken by the NPP licensee in response to the Fukushima event – Within months of the Fukushima event, NRC inspectors received guidance on how to independently assess the adequacy of actions taken by U.S. licensees in response to the Fukushima Daiichi accident. The intent of this guidance was an initial high-level look at the industry's preparedness for events that may exceed those for which the plants were designed. The inspection results were used to help evaluate the industry's readiness for a similar event and to aid in determining whether additional regulatory actions by the NRC were warranted. The inspections and associated reports identified no significant findings.

- Implementation of SAMGs by the NPP licensees – The NRC inspectors received guidance to determine the nature and extent of licensee implementation of SAMG training and exercises. The inspections and associated reports identified no significant findings.

- Compliance with mitigation strategies by the NPP licensees – The NRC inspectors received guidance to obtain verification from U.S. NPPs of their compliance with mitigation strategies associated with the loss of large areas of the plant caused by fires or explosions. To achieve these objectives the NRC issued a bulletin (Bulletin 2011-01) that required a written response from all those addressed.

The NRC staff has developed guidance for NRC inspectors to verify that U.S. fuel cycle facilities are adequately prepared to prevent and mitigate the consequences of selected beyond-design-basis events. Beyond-design-basis events are events that are possible but were not fully considered in the design process because they were judged to be too unlikely. The NRC will use the inspection results from this TI to evaluate readiness for such an event and to determine whether additional regulatory actions by the NRC are warranted. The inspection activities associated with this TI are scheduled to be completed by September 30, 2012.

4.2.1.2 Actions Taken

1. On March 18, 2011, the NRC issued IN 2011-05, "Tohoku-Taiheiyou-Oki Earthquake Effects on Japanese Nuclear Power Plants."

2. On March 23, 2011, the NRC issued TI 2515/183, "Followup to Fukushima Daiichi Nuclear Station Fuel Damage Event."

3. NRC inspectors conducted plant inspections, to assess the adequacy of actions taken by U.S. licensees in response to the Fukushima Daiichi accident. The inspections were completed on April 29, 2011, and the results were documented in NRC inspection reports. The inspection reports were issued on May 13, 2011.

4. On April 29, 2011, the NRC issued TI 2515/184, "Availability and Readiness Inspection of Severe Accident Management Guidelines."

5. NRC inspectors conducted plant inspections, to determine the nature and extent of licensees' implementation of SAMG training and exercises. They completed the inspections by May 27, 2011, and documented the results in NRC inspection reports issued by June 2, 2011.

6. On May 5, 2011, the NRC issued NRC Bulletin 2011-01, "Mitigating Strategies."

7. Licensees confirmed compliance with the bulletin within 30 days of its issuance, and provided information on their mitigation strategies programs within 60 days of the date of the bulletin.

8. On September 30, 2011, the NRC issued TI 2600/0015, "Evaluation of Licensee Strategies for the Prevention and/or Mitigation of Emergencies at Fuel Facilities."

9. In March 2012, the NRC staff completed the analysis of the licensee's responses to NRC Bulletin 2011-01.

4.2.1.3 Actions Planned

1. The NRC staff will use the analysis of the licensees' responses to NRC Bulletin 2011-01 to inform the development of guidance for the recently issued "Order Modifying Licenses with Regard to Requirements for Mitigation Strategies for Beyond-Design-Basis External Events."

2. The NRC staff will continue the inspection activities associated with TI 2600/0015.

4.2.1.4 Schedule

The staff's analysis of the licensees' responses to NRC Bulletin 2011-01 was completed in March 2012. Based on the analysis, the staff is informing the development of the guidance for the recently issued order and is in the process of determining whether further regulatory actions are needed. The staff's conclusion is expected by the summer 2012 and the guidance document by August 2012. The staff's inspection activities associated with TI 2600/0015 are scheduled to be completed by September 30, 2012. Based on the inspection findings the NRC will determine if further regulatory actions are needed.

4.2.1.5 Results

NRC inspectors independently assessed the adequacy of actions taken by U.S. licensees in response to the Fukushima Daiichi accident. They also evaluated the nature and the extent of licensee implementation of SAMGs. No significant findings were identified. A summary of observations and a results overview are available at http://www.nrc.gov/NRR/OVERSIGHT/ASSESS/follow-up-rpts.html and http://www.nrc.gov/NRR/OVERSIGHT/ASSESS/SAMGs.html.

The staff has analyzed the licensees' responses to NRC Bulletin 2011-01 and is in the process of determining whether further regulatory actions are needed. The results are not available at this point; however, the licensees' responses to the bulletin are available at http://www.nrc.gov/NRR/OVERSIGHT/ASSESS/mitigating-strategies.html.

4.2.2 Assessment Performed by the Near-Term Task Force

4.2.2.1 Discussion

In light of the accident at Fukushima Daiichi, on March 23, 2011, the NRC Chairman tasked the staff to establish a senior-level task force to conduct a systematic and methodical review of NRC processes and regulations. The NTTF was to determine whether the NRC needed to make additional improvements to its regulatory system and to provide recommendations to the Commission for its policy direction within 90 days.

The NTTF concluded that even though there was no imminent risk from continued NPP operation and licensing actives, enhancements to safety and EP were warranted. The NTTF provided 12 overarching recommendations in five general areas: (1) clarifying the regulatory framework, (2) ensuring protection, (3) enhancing mitigation, (4) strengthening EP, and (5) improving the efficiency of NRC programs.

The NTTF also determined that continued operation and continued licensing activities did not pose an imminent risk to the public health and safety of the U.S. population and environment.

4.2.2.2 Actions Taken

1. On March 23, 2011, the Chairman issued a tasking memorandum, COMGBJ-11-002, "NRC Actions Following the Events in Japan," requesting the staff to establish the NTTF.

2. On July 12, 2011, the NTTF issued its recommended safety enhancements in SECY-11-0093, "Recommendations for Enhancing Reactor Safety in the 21st Century."

3. On July 19, 2011, the NRC held a Commission briefing on the NTTF review of NRC processes and regulations following the events in Japan.

4. On July 28, 2011, the NRC held a public meeting on the NTTF's review of NRC processes and regulations following the events in Japan.

5. On August 31, 2011, the NRC held a public meeting to obtain feedback on the NTTF's recommendations.

4.2.2.3 Actions Planned

1. The NTTF has completed its actions and has published its report, as mentioned in Section 4.2.2.2. The NRC staff is now working on implementing the NTTF recommendations and the additional identified actions that should be taken in response to the Fukushima event.

4.2.2.4 Schedule

The NTTF has completed all of its actions. No further activities are scheduled for the NTTF. The JLD is overseeing the implementation of the NTTF recommendations and additional identified actions.

4.2.2.5 Results

The NTTF issued its recommended safety enhancements in SECY-11-0093, "Recommendations for Enhancing Reactor Safety in the 21st Century."

4.2.3 The NRC Staff's Review and Implementation of the NTTF Recommendations

4.2.3.1 Discussion

The Commission directed the staff to promptly engage with stakeholders to review and assess the recommendations of the NTTF in a comprehensive and holistic manner. The purpose of this assessment was to provide the Commission with fully informed options and recommendations. The Commission instructed the staff to remain open to strategies and proposals presented by stakeholders, expert staff members, and others as it provided its recommendations to the Commission. The staff was directed to develop a paper, within 21 days of the issuance of the NTTF report, reviewing the actions the NTTF recommended be taken without unnecessary delay (see SECY-11-0124). The Commission also directed the staff to submit a second paper within 45 days of the issuance of the NTTF report, prioritizing all recommended actions to be taken in response to the lessons learned from Fukushima (see SECY-11-0137). Prioritization of the recommendations was subject to the review and approval of the Steering Committee.

The near-term actions that the staff identified in SECY-11-0124 which should be taken without unnecessary delay focused on:

*   seismic and flood hazard reevaluations

*   seismic and flooding walkdowns

*   SBO regulatory actions

*   equipment needed to recover from the loss of large areas due to explosions or fire

*   reliable hardened vents for Mark I containments

*   strengthening and integrating EOPs, SAMGs, and EDMGs

*   EP regulatory actions

The NRC staff developed a SECY paper that used the following method for prioritizing all NTTF recommendations:

- Reflect regulatory actions to be taken by the staff in response to the Fukushima lessons learned.

- Identify implementation challenges.

- Include the technical and regulatory bases for the prioritization.

- Identify additional recommendations, if any.

- Include a schedule and milestones with recommendations for appropriate stakeholder engagement.

In some cases, subsequent recommendations from the Steering Committee have modified or enhanced the NTTF recommendations.

As a result of the staff's assessment and prioritization by the Steering Committee, the NTTF recommendations were categorized in SECY-11-0137 into three tiers:

- Tier 1 - consists of those recommendations that should be started without delay.

- Tier 2 - consists of those recommendations that could not be initiated in the near-term because they need further technical assessment, depend on Tier 1 issues, or require the availability of critical skill sets.

- Tier 3 - consists of those recommendations that require further staff study to support a regulatory action.

4.2.3.2 Actions Taken

1. On August 19, 2011, the Commission issued the SRM on SECY-11-0093, "Near-Term Report and Recommendations for Agency Actions Following the Events in Japan."

2. On September 9, 2011, the NRC staff issued SECY-11-0124, "Recommended Actions to Be Taken Without Delay from the Near-Term Task Force Report."

3. On October 3, 2011, the NRC staff issued SECY 11-0137, "Prioritization of Recommended Actions to be taken in Response to Fukushima Lessons Learned."

4. On October 18, 2011, the Commission issued the SRM on SECY-11-0124, "Recommended Actions to be taken Without Delay from the Near-Term Task Force Report."

5. On December 15, 2011, the Commission issued the SRM on SECY-11-0137, "Prioritization of Recommended Actions to be taken in Response to Fukushima Lessons Learned."

6. On February 17, 2012, the NRC issued SECY-12-0025 to provide (1) the proposed orders; (2) the proposed RFIs; (3) the resolution of the six additional issues from

SECY-11-0137; and (4) the resolution of the Advisory Committee on Reactor Safety recommendations.

7. On March 9, 2012, the Commission approved actions in SECY-12-0025.

8. On March 12, 2012, the NRC staff issued the orders and RFIs.

4.2.3.3 Actions Planned

1. The NRC staff will evaluate additional recommendations which may be related to lessons-learned from the Fukushima accident as they are identified. As appropriate, these recommendations will be prioritized and addressed.

4.2.3.4 Schedule

The NRC staff will evaluate additional recommendations as they are identified.

4.2.3.5 Results

The NRC has determined that continued operation and continued licensing activities associated with U.S. NPPs do not pose an imminent risk to the health and safety of the U.S. population and environment. The NTTF concluded, however, that even though there was no imminent risk from continued NPP operation and licensing actives, enhancements to safety and EP were warranted. In light of these needed enhancements, the NRC staff continues to evaluate all of the Fukushima-related recommendations to assess them in a comprehensive and holistic manner.

4.2.4 Establishment of the Japan Lessons-Learned Project Directorate

4.2.4.1 Discussion

As a result of Commission direction related to SECY-11-0117, which established a proposed charter for future Japan-related work, the JLD was established to oversee implementation of the NTTF's recommendations, identify and evaluate additional recommendations, and oversee the development of plans to address the items identified as requiring long-term review. The JLD works in conjunction with the other NRC offices to achieve these goals.

The JLD is guided by a steering committee composed of senior NRC managers who report directly to the EDO. The Steering Committee is responsible for providing high-level direction and management vision in light of the events that took place in Japan.

4.2.4.2 Actions Taken

1. On August 26, 2011, the NRC staff issued SECY 11-0117, "Proposed Charter for the Longer-Term Review of Lessons Learned from the March 11, 2011, Japanese Earthquake and Tsunami."

2. On October 19, 2011, the Commission issued the SRM on SECY-11-0117, "Proposed Charter for the Longer-Term Review of Lessons Learned from the March 11, 2011, Japanese Earthquake and Tsunami."

3. On November 7, 2011, the JLD was established as part of NRR.

4.2.4.3 Actions Planned

1. The JLD, as well as the technical staff, will coordinate the implementation of all the actions in response to the Fukushima event.

4.2.4.4 Schedule

The schedule for the implementation of the actions in response to the Fukushima event is discussed throughout this report.

4.2.4.5 Results

The activities are currently ongoing and results are not yet available.

4.3 Interactions with Congress and Other Government Agencies

4.3.1 Discussion

After receiving news of the Japan earthquake, on March 11, 2011, the NRC's Office of Congressional Affairs (OCA) started staffing the Headquarters operations center (HOC) 24 hours a day and provided information to congressional staffers (including press releases, and responses to telephone and e-mail requests for information). OCA also participated in numerous briefings with other Federal agencies.

The NRC's OCA staff supported requests for information about the Japan event, including coordinating twice daily telephone briefings to interested congressional staffers looking for up-to-date information on the situation in Japan. To supplement the staffing levels needed to perform these actions, OCA hired a retired civil servant with years of OCA experience.

The OCA Director accompanied the NRC Chairman to Congress to discuss various aspects of the situation in Japan and the NRC's response.

Briefings and meetings with congressional stakeholders are key parts of OCA's mission. By engaging with congressional stakeholders, the NRC maintains an open and cooperative environment that enhances Congress' ability to enact appropriate measures. In support of this goal, many congressional briefings (in person and by telephone with members of Congress and congressional staffers) have been conducted to respond to requests for information.

4.3.2 Actions Taken

1. Provided staff to perform liaison duties in the NRC Operations Center. OCA has eight staffers who are trained to respond to such an event, and have participated in many drills over the years to ensure they are able to respond appropriately.

2. Developed (and continues to maintain) a distribution list of more than 600 congressional points of contact to five updated information about the NRC's response to the events in Japan, the NTTF's recommendations and Commission decisions.

3. Provided all relevant information concerning the situation in Japan to the NRC's oversight committees, congressional delegations, and any other members who requested information.

4. Participated in several joint congressional briefings with the NRC's Federal partners, including DOE, DOS, DHS, NOAA, EPA, and DOD.

5. Since the Fukushima events unfolded, the NRC has testified at over 20 hearings, most of which covered the situation in Japan and the actions the NRC has taken in response.

6. Participated in twice-daily U.S. Agency for International Development telephone briefings to congressional staff members to provide up-to-date information on the situation in Japan, and held NRC-specific briefings for congressional staffers.

7. Responded to numerous e-mails and telephone calls from congressional staffers about issues associated with the situation in Japan and its potential impact to the U.S.

8. Provided a "year-end" briefing to interested congressional staffers to describe what the NRC has done in 2011, including an update on actions associated with the NRC's response to Fukushima.

4.3.3 Actions Planned

1. Continue to give relevant information to Congress as it becomes available to ensure Congress is currently and fully informed.

2. Maintain an e-mail distribution list for updates related to the NTTF's recommendations.

4.3.4 Schedule

NRC will update Congress, as appropriate.

4.3.5 Results

The NRC has regularly engaged with congressional stakeholders to maintain an open and cooperative environment that enhances Congress' ability to enact appropriate measures. The activities related to Fukushima are currently ongoing and results are not yet available.

4.4 Interactions with the Nuclear Industry, Members of the Public, and Stakeholders

4.4.1 Discussion

As a result of the Japan event, the NRC has interacted with the U.S. nuclear industry, members of the public, and other stakeholders on how to best implement the actions contained in the NTTF report and additional Commission direction. A kickoff meeting on the implementation of near-term recommendations was held on December 1, 2011, between the NRC staff, representatives from NEI and nuclear industry, and members of the public. During this meeting, the NRC staff and the participating stakeholders discussed how to most efficiently move forward in implementing the NTTF near-term recommendations. Subsequently, the NRC planned and held public meetings for each of the near-term recommendations that required the issuance of an order or RFI. These meetings were instrumental in obtaining early input which was factored into the specific implementation plans for all the issues. These meetings were highly effective in engaging external stakeholders in an open and transparent manner.

4.4.2 Actions Taken

1. The NRC held a public meeting on October 7, 2011, and February 9, 2012, with the Advisory Committee on Reactor Safeguards.

2. The Commission held a meeting on November 29, 2011, with the Advisory Committee on Reactor Safeguards.

3. The NRC held a public meeting on December 1, 2011, with members of the public, NEI, and industry, to discuss strategies to implement the NTTF near-term recommendations.

4. The NRC held public meetings on December 8, 2011, January 18, March 28, and April 24, 2012, with members of the public, NEI, and industry, to discuss actions related to NTTF Recommendation 4.2, Mitigating Strategies.

5. The NRC held public meetings on December 12, 2011, January 9, January 19, 2012, February 7, and March 5, 2012, with members of the public, NEI, and industry, to discuss actions related to NTTF Recommendation 9.3, EP Staffing Analysis and Communications.

6. The NRC held public meetings on December 14, 2011, January 18, February 22, March 1-2, March 15, March 27, April 2-3, April 11-12, April 13, April 17, April 25-26, and April 27, 2012, with members of the public, NEI, and industry, to discuss actions related to NTTF Recommendations 2.1 and/or 2.3, Flooding and Seismic Protections.

7. The NRC held public meetings on December 15, 2011, January 17, and May 2, 2012, with members of the public, NEI, and industry, to discuss actions related to NTTF Recommendation 5.1, Reliable Hardened Vents.

8. The NRC held public meetings on December 15, 2011, January 19, and March 29, April 18, and May 1, 2012, with members of the public, NEI, and industry, to discuss actions related to NTTF Recommendation 7.1, SFP Instrumentation.

9. On January 13, 2012, the NRC Steering Committee met with representatives from NEI and the nuclear industry Committee to discuss the path forward on the implementation near-term recommendations.

10. On March 12, 2012, the NRC staff issued orders and RFIs to all licensees. These orders and RFIs are discussed in SECY-12-0025.

4.4.3 Actions Planned

1. The NRC staff will continue to engage with stakeholders to support associated guidance development, as necessary.

4.4.4 Schedule

The schedule for the implementation of the actions in response to the Fukushima event is discussed throughout this report. The NRC staff will, however, continue to engage with stakeholders to support associated guidance development, as necessary.

4.4.5 Results

The NRC has determined that continued operation and continued licensing activities associated with U.S. NPPs do not pose an imminent risk to the public health and safety of the U.S. population and environment. The NTTF concluded, however, that even though there was no imminent risk from continued NPP operation and licensing actives, enhancements to safety and EP were warranted. In light of these needed enhancements, the NRC staff continues to evaluate all of the Fukushima-related recommendations to assess and disposition them in a comprehensive and holistic manner.

TOPIC 5. EMERGENCY PREPAREDNESS AND RESPONSE AND POST-ACCIDENT MANAGEMENT

5.1 Introduction

The accident at the Fukushima Daiichi NPP demonstrated the need for effective post-accident management, including radiological evaluation, efficient mechanisms for decisionmaking, control and management of contaminated goods, resettlement, communication and information, remediation activities, and indemnifications.

The NRC determined that the accident at Fukushima highlights a need to evaluate EP activities when addressing beyond-design-basis accidents. This section of the report will discuss NRC post-Fukushima actions in the context of multi-unit events and prolonged SBO, incident response and EP, communications, and radiation protection. The EP framework was recently changed to reflect technological advances and address a number of security enhancements that resulted from the events of September 11, 2001. The revised EP rule, which was in effect on December 23, 2011, directly correlates to certain post-Fukushima EP safety enhancements. The rule can be found on www.regulations.gov, Docket ID: NRC-2008-0122, or at http://www.gpo.gov/fdsys/browse/collection.action?collectionCode=FR, (FR DOC# 2011-29735 or 76 FR 72560). Below are highlights of pertinent U.S. EP regulations.

- Staffing - The revised EP rule amends 10 CFR Part 50, Appendix E, Section IV(A)(9), to address staffing concerns regarding the assignment of responsibilities of the on-shift personnel, such that during a response to an emergency event the on-shift personnel are not overburdened. Before this revision, the regulations required "adequate" on-shift staffing levels, and guidance documents (such as Interim Staff Guidance NSIR/DPR-ISG-01, "Emergency Planning for Nuclear Power Plants," dated November 2011), more clearly defined the meaning of "adequate." The revised EP rule requires licensees to conduct a detailed analysis demonstrating that on-shift personnel assigned to implement emergency plans are not given responsibilities that would prevent the timely performance of their assigned functions as specified in the emergency plan.

- Dose Assessment - Under 10 CFR 50.47(b)(9), the NRC requires licensees to have a means for assessing the potential offsite consequences of a radiological emergency. Appendix E, "Emergency Planning and Preparedness for Production and Utilization Facilities," to 10 CFR Part 50, describes requirements for emergency planning and preparedness for production and utilization facilities in the U.S, including the requirements of each facility's emergency plans. Accordingly, a requirement of the emergency plan includes assessing the impact of a radiological release and providing a protective action recommendation.

- Communication - After an emergency is declared, a licensee must first notify the State and local response organizations, and then the NRC. The NRC requires that licensees provide these emergency notifications with accurate and timely data on: (1) core and coolant system conditions, (2) conditions inside the containment building, (3) radioactivity release rates, and (4) data from the plant's meteorological tower. Regulations at 10 CFR 50.47(b)(6) and Appendix E to 10 CFR Part 50 provide for these notifications and clearly specify the times in which licensees must make the declarations, as well as the methods by which they must be made. The NRC also uses the Emergency Response Data System (ERDS), which allows for the direct electronic transmission of selected NPP parameters. Regulations at 10 CFR 50.72, "Immediate Notification Requirements for Operating Nuclear Power Plants," and Appendix E to 10 CFR Part 50, require power reactor licensees to transmit ERDS data to the NRC when an emergency has been declared. Memoranda of understanding exist between neighboring countries so that their governments can be well informed if there is an event at a U.S. facility.

To meet the U.S. commitment under the International Atomic Energy Agency (IAEA) Convention on Early Notification of a Nuclear Accident, the NRC will promptly notify IAEA if a serious accident occurs at a commercial NPP. In addition to providing updated technical reports, NRC will work with DOS and other U.S. agencies to provide additional information to the IAEA.

- Drills and Exercises - Under 10 CFR 50.47(b)(14), the NRC requires licensees to periodically conduct drills and exercises to develop key skills and evaluate major portions of their emergency response capabilities. Regulations at 10 CFR 50.47(b)(15) provides requirements for radiological emergency response training, and NUREG-0654, "Criteria for Preparation and Evaluation of Radiological Emergency Response Plans and Preparedness in Support of Nuclear Power Plants," dated November 1980, includes further details on implementing these regulations. The revised EP rule added requirements for more challenging drills and exercises, as well as hostile-action-based exercises.

5.2 Multi-Unit Events and Station Blackouts - Staffing and Communication

5.2.1 Identify Staff Needed for Responding to a Multi-Unit Event

5.2.1.1 *Discussion*

In the U.S., NRC licensees' EP plans must ensure adequate staffing of emergency response personnel (onsite and offsite) to effectively respond to an emergency. As a result of the events at Fukushima Daiichi, the NRC is evaluating whether more regulatory actions regarding a licensee's ability to responds to multi-unit events may be warranted. The NRC issued an RFI to licensees to determine whether additional changes to NRC regulations or guidance documents are necessary to ensure appropriate staffing to respond to a multi-unit event. The NRC staff will evaluate the responses to determine appropriate regulatory actions, if any (NTTF Recommendation 9.3). Specific actions to address this are included below.

5.2.1.2 *Actions Taken*

1. On October 3, 2011, the NRC staff proposed that the NRC prioritize the NTTF recommendations to be taken in response to the lessons learned from Fukushima (SECY-11-0137). On December 15, 2011, the Commission approved the staff's proposed prioritization of the NTTF recommendations (SRM on SECY-11-0137).

2. The NRC staff has engaged stakeholders to inform the development of a methodology to perform a staffing study to determine the required staff to fill all necessary positions to respond to a multi-unit event.

3. On March 12, 2012, the NRC issued an RFI to licensees to (1) perform a staffing study to determine the required staff to fill all necessary positions to respond to a multi-unit event, and (2) inform the NRC of the results of the staffing study and any actions taken or planned, along with their implementation schedules. (NTTF Recommendation 9.3). This RFI is discussed in SECY-12-0025.

5.2.1.3 *Actions Planned*

1. The NRC will evaluate licensee responses and take regulatory action to require implementation, as appropriate.

5.2.1.4 *Schedule*

The staff will evaluate licensee responses to RFIs and take regulatory action, as appropriate.

5.2.1.5 *Results*

The activities are currently ongoing and results are not yet available.

5.2.2 Periodic Training and Exercises for Multi-Unit and Prolonged SBO Scenarios

5.2.2.1 *Discussion*

In the U.S., EP must ensure adequate training of emergency response personnel (onsite and offsite) so licensees can effectively respond to emergency events. As a result of the events at Fukushima Daiichi, additional regulatory actions may be needed to determine whether the current requirements for training are adequate (NTTF Recommendation 9.1 and 9.2). Specific actions to address this are included below.

5.2.2.2 *Actions Taken*

1. On October 3, 2011, the NRC staff proposed that the NRC prioritize the NTTF recommendations to be taken in response to the lessons learned from Fukushima (SECY-11-0137). On December 15, 2011, the Commission approved the staff's proposed prioritization of the NTTF recommendations (SRM on SECY-11-0137).

5.2.2.3 *Actions Planned*

1. The NRC will evaluate and determine whether a rulemaking is warranted to address the need for licensees to conduct periodic training and exercise for multi-unit and prolonged SBO scenarios. (NTTF Recommendations 9.1 and 9.2)

5.2.2.4 *Schedule*

This is a long-term action and a detailed schedule has not been developed.

5.2.2.5 *Results*

This is a long-term effort and results are not yet available.

5.2.3 Onsite and Offsite Communications

5.2.3.1 *Discussion*

The Fukushima incident highlighted the need to provide a way to power communications equipment needed to communicate onsite (e.g., radios for response teams and communication between facilities) and offsite (e.g., cellular telephones, satellite telephones) during a prolonged SBO (NTTF Recommendation 9.3). Specific actions to address this are included below.

5.2.3.2 *Actions Taken*

1. On October 3, 2011, the NRC staff proposed that the NRC prioritize the NTTF recommendations to be taken in response to the lessons learned from Fukushima (SECY-11-0137). On December 15, 2011, the Commission approved the staff's proposed prioritization of the NTTF recommendations (SRM on SECY-11-0137).

2. The NRC staff has engaged with stakeholders to discuss potential enhancements to communications systems and power supplies to ensure a means to power communications equipment necessary for licensee onsite and offsite communications is maintained during a prolonged SBO event.

3. On March 12, 2012, the NRC issued an RFI to licensees to (1) conduct an assessment of the communications systems and power supplies to determine what enhancements would be needed to power communications systems and equipment necessary for licensee onsite and offsite communications during a prolonged SBO event, and (2) inform the NRC of the results of the study of communications systems and power supplies and any actions taken or planned, to enhance the communication systems, along with their implementation schedules. (NTTF Recommendation 9.3) This RFI is discussed in SECY-12-0025.

5.2.3.3 *Actions Planned*

1. The NRC staff will evaluate licensee responses to the RFI and take appropriate regulatory action to require implementation.

5.2.3.4 *Schedule*

The NRC staff will evaluate licensee responses to the RFI and take regulatory action, as appropriate.

5.2.3.5 *Results*

The activities are currently ongoing and results are not yet available.

5.3 Maintain Emergency Response Data System Capability

5.3.1 Discussion

The Fukushima incident highlighted the need for accurate, real-time data from the site. As part of the current effort to modernize the ERDS infrastructure, the NRC has developed a replacement to the existing modems. The current regulatory requirements do not require that

ERDS data be available when power supplies are lost and transmission capability may be affected. Another challenge evident from Fukushima is the need for archived data to aid in the reconstruction of events after an accident (NTTF Recommendations 9.3 and 9.4).

5.3.2 Actions Taken

1. The NRC is currently completing the ERDS virtual private network transition that has upgraded NRC servers, upgraded user interfaces, allowed Web access, and replaced obsolete analog modem technology.

5.3.3 Actions Planned

1. The NRC will complete the transition to the ERDS virtual private network.

2. The NRC staff will evaluate additional regulatory options to address redundant transmission capability, completeness of data, and continuous monitoring capability for ERDS information. (NTTF Recommendations 9.3 and 9.4)

5.3.4 Schedule

The transition to the virtual private network is ongoing and expected to be complete in June 2012. Additional regulatory actions will require that the NRC conduct a long-term study and coordinate with the ongoing effort to modernize ERDS. A schedule for this long-term study is not currently available.

5.3.5 Results

The activities are currently ongoing and results are not yet available.

5.4 Radiation Protection

5.4.1 Discussion

The accident at Fukushima presents challenges with respect to dose assessment capability because of the multi-unit nature of the release. The presence of extended releases from multiple units and SFPs at Fukushima has highlighted the need to have the ability to model multi-unit accidents (NTTF Recommendation 9.3).

5.4.2 Actions Taken

1. On October 3, 2011, the NRC staff proposed that the NRC prioritize the NTTF recommendations to be taken in response to the lessons learned from Fukushima (SECY-11-0137). On December 15, 2011, the Commission approved the staff's proposed prioritization of the NTTF recommendations (SRM on SECY-11-0137).

5.4.3 Actions Planned

1. The NRC staff will evaluate additional regulatory options that may be needed to address how licensees should add guidance to their emergency plans that documents how to perform a multi-unit dose assessment (including releases from SFPs) using a licensee site-specific dose assessment software and approach.

5.4.4 Schedule

This is considered a long-term action and a schedule will be developed in the summer 2012.

5.4.5 Results

This is a long-term action and results are not yet available.

5.5 Pre-Staging of Potassium Iodide Beyond the 10-Mile Radius

5.5.1 Discussion

The NRC staff has determined that the current 10-mile potassium iodine (KI) distribution zone remains adequate. However, the NRC staff will continue to monitor and evaluate population health studies to confirm this determination.

5.5.2 Actions Taken

1. The NRC has determined that the current 10-mile KI distribution zone remains adequate.

5.5.3 Actions Planned

1. The staff intends to continue to study the health effects on populations around NPPs. The staff will also continue to monitor and evaluate the population health studies that have been proposed by the Japanese Government.

5.5.4 Schedule

This is considered a long-term effort and a schedule will be developed in the summer 2012.

5.5.5 Results

This is a long-term action and results are not yet available.

5.6 Communication, Transparency, and Openness

5.6.1 Discussion

After receiving news of the Japan earthquake, on March 11, 2011, the NRC staffed its HOC and disseminated the first of many press releases and NRC blog posts.

The NRC's Office of Public Affairs (OPA) staff provided extended coverage for several months, to support requests for information about the Japan event. All media inquiries of the NRC HOC were sent to a media desk, where they were recorded, prioritized, and sent to public affairs officers at Headquarters for response. OPA staff members from the regional offices handled local media inquiries. OPA also brought in more than a dozen employees from around the NRC to augment the six professionals who comprise the Headquarters public affairs staff and allow OPA to respond to hundreds of requests for information from domestic and international media outlets and the public. Two public affairs staffers from FEMA also came to support the NRC's response efforts for several days.

To respond to an unprecedented level of public telephone calls and e-mails, OPA created the NRC's first-ever public inquiry desk. This desk answered telephone and e-mail inquiries for extended daily hours and on weekends for several months. A special page was created on the NRC Web site to make it easier for the public to obtain Japan-specific information, including press releases, questions and answers documents, background information and fact sheets, and links to the NRC blog posts.

The OPA Director accompanied the NRC Chairman to the White House and Congress. Coverage of the NRC's role in the overall U.S. Government's response to the events in Japan and how the Japanese incident related to the safety of U.S. NPPs was intense and continuous. A media monitoring contractor provided extensive reports on press coverage.

Meeting with stakeholders is a key portion of each step in the process for all EP recommendations related to the Fukushima lessons learned project. By engaging with stakeholders, the NRC will maintain a collaborative environment that will enhance the project's results and keep the public informed. All EP actions related to the lessons learned from the Fukushima Daiichi event and the resulting impact on U.S. nuclear facilities are posted on the NRC's public Web site and, as appropriate, included in periodic press releases.

The NRC held a series of public meetings on the EP recommendations to directly engage stakeholders and industry. The dates of these meetings were December 12, 2011, January 9, January 19, February 7, and March 5, 2012; more will follow, if needed.

5.6.2 Actions Taken

1. The NRC updated its crisis communication plan with lessons learned and added staff to the OPA's technical briefer list to support public and media outreach efforts in future events.

2. The NRC developed a standard operating protocol using lessons learned from the implementation of the public inquiry desk; the agency is developing a permanent roster.

3. The NRC held public meetings to discuss NTTF recommendations and to obtain input from the public and interested stakeholders on how to address the recommendations of the NTTF that relate to EP lessons learned from the Fukushima response.

4. The NRC conducted a post-event response After Action Review (AAR) that addressed needed enhancements to our real-time communications with stakeholders, such as the States, during events.

5.6.3 Actions Planned

1. Develop a roster to staff the public inquiry desk during future events.

2. The NRC is participating in a Federal Government working group to update the Emergency Support Function 15 – External Communication chapter and annexes to the U.S. "National Response Framework," dated January 2008, which will include a new section addressing communications response to international incidents as well as updating expected coordinated responses to domestic incidents.

3. Resolve the items identified in the AAR report.

5.6.4 Schedule

The schedule for developing the roster to staff the inquiry desk has not been determined. The updated National Response Framework Emergency Support Function 15 annexes are scheduled to be delivered to the U.S. President before the end of 2012. The AAR action items are currently being prioritized in near- and long-term activities. A detailed schedule for completing the AAR action items is not currently available.

5.6.5 Results

This is a long-term action and results are not yet available.

5.7 Additional Long-Term EP-Related Topics

5.7.1 Discussion

The NTTF report identified additional long-term EP-related areas that need further evaluation and enhancement. Specifically, the NTTF recommended that the NRC staff review three additional EP-related topics associated with multi-unit events and prolonged SBOs; as well as additional topics related to decisionmaking, radiation monitoring, and public education. (NTTF Recommendations 10 and 11)

5.7.2 Actions Taken

1. No specific actions have been taken on these topics.

5.7.3 Actions Planned

1. The NRC is currently evaluating the following topics related to EP:

   - Analyze current protective equipment requirements for emergency responders and guidance based upon insights from the accident at Fukushima. (NTTF Recommendation 10.1)

   - Evaluate the command and control structure and the qualifications of decisionmakers to ensure that the proper level of authority and oversight exists in the correct facility for a long-term SBO or multi-unit event accident or both. (NTTF Recommendation 10.2)

   - Evaluate ERDS to determine an alternate method for transmittal, and whether the current data set is sufficient. (NTTF Recommendation 10.3)

   - Study whether enhanced onsite emergency response resources are necessary to support the effective implementation of the licensees' emergency plans, including the ability to deliver the equipment to the site under conditions involving significant natural events where degradation of offsite infrastructure or competing priorities for response resources could delay or prevent the arrival of offsite aid. (NTTF Recommendation 11.1)

- Work with FEMA, State officials, and other external stakeholders to evaluate insights from the implementation of EP at Fukushima to identify potential enhancements to the U.S. decisionmaking framework, including the concepts of site recovery and re-entry. (NTTF Recommendation 11.2)

- Study the efficacy of real-time radiation monitoring onsite and within the emergency planning zones (including consideration of AC independent and real-time availability on the internet). (NTTF Recommendation 11.3)

- Conduct training, in coordination with the appropriate Federal partners, on radiation, radiation safety, and the appropriate use of KI in the local community around each NPP. (NTTF Recommendation 11.4)

2. The NRC staff will write a paper to the NRC Commission describing the plans and schedules for performing work on these additional topics.

5.7.4 Schedule

A Commission paper describing the plans and schedules for performing work on these additional topics is scheduled to be provided to the Commission in July 2012.

5.7.5 Results

This is a long-term action and results are not yet available.

TOPIC 6. INTERNATIONAL COOPERATION

6.1 Introduction

The accident at Fukushima highlights the importance of continuous international cooperation. Since the accident occurred, the U.S. Government has augmented its coordinated program of international nuclear safety activities. Some existing activities have been expanded to address lessons learned from the accident, while certain initiatives have been created specifically to address the accident and its implications. In both cases, the objectives of U.S. representatives in international meetings have been to expand their understanding of the accident and others' approaches to learning its lessons; to share relevant experience and lessons learned; and to minimize duplication of effort and leverage financial and human resources. In addition to contributing to the U.S. Government's direct cooperation with and support of the Government of Japan, the NRC has actively supported numerous post-Fukushima international activities, both on a bilateral and a multilateral basis. Through regular communication with its foreign government counterparts, particularly in the regulatory area, and participation in international meetings, the U.S. has gleaned valuable information to enhance its domestic nuclear safety program and has contributed to the development of a stronger global nuclear safety regime.

Within its areas of responsibility, the NRC supports the U.S. foreign policy goal of ensuring the safe and secure use of nuclear and radioactive materials and guarding against the spread of nuclear weapons. As explained in greater detail in the "United States of America Fifth National Report for the Convention on Nuclear Safety," dated September 2010, the NRC actively participates in developing and implementing a variety of legally binding treaties and conventions that create an international framework for the peaceful uses of nuclear energy. The NRC provides technical and legal advice and assistance to international organizations and foreign countries as they work to develop effective regulatory organizations and rigorous safety standards. Some activities are carried out within the programs of the IAEA, the Nuclear Energy Agency (NEA) of the Organization for Economic Co-operation and Development (OECD), or other international organizations. The NRC conducts other activities directly with counterpart agencies in other countries under bilateral cooperation agreements.

6.2 International Treaties

6.2.1 Discussion

Among the treaties that legally bind the U.S. Government's peaceful uses of nuclear energy and nuclear applications are the 1978 Nuclear Non-Proliferation Treaty, the 1980 Convention on Physical Protection of Nuclear Material, the 1994 Convention on Nuclear Safety (CNS), the 1986 Convention on Early Notification of a Nuclear Accident, the 1986 Convention on Assistance in Case of a Nuclear Accident or Radiological Emergency, and the 1997 Joint Convention on the Safety of Spent Fuel Management and on the Safety of Radioactive Waste Management.

6.2.2 Actions Taken

1. The NRC maintained leadership roles (Vice President and Country Group Vice Chair) in the CNS review process.

2. The NRC participated in U.S. Government discussions with international counterparts on proposals to either amend relevant conventions or enhance their effectiveness through procedural means.

3. The NRC participated in the international negotiation of the IAEA "Action Plan on Nuclear Safety," which was approved by the IAEA Board of Governors and endorsed by the IAEA General Conference in September 2011.

4. The NRC informed the IAEA that in April 2011, it had appointed a National Competent Authority contact at the NRC for the Early Convention and Assistant Convention.

5. The NRC submitted 2 proposals to enhance the CNS effectiveness on April 19, 2012.

6. The NRC is supporting the U.S. delegation to the first Preparatory Committee meeting of Parties to the Non-Proliferation Treaty.

7. The NRC is supporting the U.S. Government in its development of implementing legislation to complete its ratification of the amendment to the Convention on Physical Protection of Nuclear Material.

6.2.3 Actions Planned

1. Participate in the Joint Convention review meeting. The NRC will co-deliver the U.S. National Report, will chair a Country Group and will assist the U.S. delegation during the two-week review meeting.

2. Participate in the CNS extraordinary meeting. The NRC will maintain its leadership roles (i.e., Vice President and Country Group Vice Chair) and will assist the U.S. delegation during the one-week meeting.

3. Participate in the sixth CNS review meeting.

4. Support the Non-Proliferation Treaty Preparatory Committee meetings and the Review Conference.

6.2.4 Schedule

U.S. representatives will participate in the Joint Convention review meeting, which will take place in May 2012, in Vienna, Austria. The U.S. will also attend the CNS extraordinary meeting in August 2012, and the sixth CNS review meeting in April 2014, both in Vienna, Austria. U.S. representatives will also attend the Non-Proliferation Treaty Preparatory Committee meetings in 2013 and 2014, and the Review Conference in 2015.

6.2.5 Results

Participation in these ongoing activities has provided the U.S. with multiple opportunities to benefit from the peer review process set forth by the various international instruments. It has also enabled the U.S. to share experience and insights with the international community as nuclear safety-related documents and proposals have been crafted.

6.3 Communications

6.3.1 Discussion

The NRC supports U.S. Government activities to improve communications before, during, and after a nuclear or radiological event. Through its treaty and convention obligations, the U.S. Government participates in international emergency response activities such as the emergency response trials, drills, and IAEA Incident and Emergency Center convention exercises (ConvEx); provides funding and staff support for international EP, including the IAEA's International Emergency Center and the NEA's Working Party on EP; and routinely coordinates with neighboring States on natural disaster and other emergency communication events.

To meet the U.S. commitment under the IAEA Convention on Early Notification of a Nuclear Accident, the NRC will promptly notify IAEA if a serious accident occurs at a commercial NPP. In addition to providing updated technical reports, the NRC will work with DOS and other U.S. Government agencies to provide additional information to the IAEA.

Since 1992, the U.S. has participated in the International Nuclear and Radiological Event Scale (INES) system that the IAEA and NEA have established as a means for promptly communicating, to the public and media, the safety significance of events that may occur at nuclear or radiological facilities. In 2002, NRC expanded its participation in INES to submit ratings for those events rated at Level 2 or higher on INES. The U.S. has also played a significant role on IAEA's INES Advisory Committee, including support of the international negotiations that resulted in the INES and events related to its expanded use for rating radiation and transport events.

The NRC has agreements with its neighboring states, Canada and Mexico, to promptly notify and exchange information in the event of an emergency that has the potential for transboundary effects. The agreement with Canada, "Agreement Between the Government of the United States of America and the Government of Canada on Cooperation in Comprehensive Civil Emergency Planning and Management," is implemented by the procedure specified in "Administrative Arrangement Between the United States Nuclear Regulatory Commission and the Atomic Energy Control Board of Canada for Cooperation and the Exchange of Information in Nuclear Regulatory Matters," both dated June 21, 1989. The agreement between the NRC and the Canadian Nuclear Safety Commission (CNSC) was most recently renewed in 2007. The agreement with Mexico, "Agreement for the Exchange of Information and Cooperation in Nuclear Safety Matters," is implemented by the "Implementing Procedure for the Exchange of Technical Information and Cooperation in Nuclear Safety Matters Between the Nuclear Regulatory Commission of the United States of America and the Comisión Nacional de Seguridad Nuclear y Salvaguardias of Mexico (CNSNS)," both dated October 6, 1989. This agreement was most recently renewed in 2007.

After the events at Fukushima, recognizing the proximity of several U.S. NPPs to the borders with Canada and Mexico, the NRC engaged in close communications with the neighboring countries.

6.3.2 Actions Taken

1. The NRC and CNSC continued to exchange information regularly as details about the Fukushima accident unfolded, in part because of shared public concerns about radiation

affecting the western coasts of Canada and the U.S. While such discussions with Mexico were not as immediately necessary, the U.S. Government continued to exchange information with its Mexican counterparts in the areas of EP and nuclear reactor safety.

2. The NRC and CNSC held a technical bilateral meeting to exchange information on a variety of subjects, including enhancing bilateral EP cooperation and sharing experience and lessons learned to strengthen both domestic programs in February 2012.

3. The NRC held a planning meeting with CNSNS in March 2012 to discuss future cooperation meetings and technical bilateral exchanges in 2012.

6.3.3 Actions Planned

1. The U.S. will continue to maintain close cooperative relationships with its North American neighbors. This will include sharing information about the nuclear safety assessment activities currently underway at U.S. facilities and discussing ways to enhance coordination in the EP area.

2. The NRC will schedule additional technical bilateral meetings with CNSC and CNSNS in 2012.

6.3.4 Schedule

Bilateral activities will be scheduled on an as-needed basis. The U.S. Government will also continue to interact with both Canada and Mexico on a regular basis in multilateral fora and in peer review forums such as the CNS review meetings and the extraordinary meeting.

6.3.5 Results

Interactions with Canadian officials in the short-term and long-term aftermath of the Fukushima accident have been beneficial in sharing data and approaches. Participation in international EP exercises such as the ConVex exercise hosted by Mexico has helped ensure that proper procedures are in place for robust cooperation in the event of an accident affecting North American borders. A Commissioner-level visit to Mexico in February 2012 provided an opportunity to discuss Fukushima lessons learned and likely changes in each country's regulatory program.

6.4 Bilateral and Multilateral Cooperation and Assistance Program

6.4.1 Discussion

The NRC has arrangements to exchange technical information with nuclear safety agencies in more than 41 countries and Taiwan. In addition, the NRC works with many countries that are considering the option of nuclear power through the International Regulatory Development Partnership, although there may not yet be a formal bilateral arrangement in place. The NRC and its foreign counterparts routinely exchange operational safety data and other regulatory information. The NRC provides safety, security, EP, and safeguards advice, training, and other assistance to countries that seek U.S. help to improve their regulatory programs.

The NRC's information exchange arrangements serve as communication channels with foreign regulatory authorities, establishing a framework for the NRC to gain access to non-U.S. safety information that can (1) alert the U.S. Government and industry to potential safety problems, (2) help find possible accident precursors, and (3) provide accident and incident analyses, including lessons learned, that could be directly applicable to the safety of U.S. NPPs and other facilities. The arrangements also serve as a vehicle for the assistance the NRC provides to countries to establish and improve their regulatory capabilities and infrastructure. Thus, the arrangements facilitate the NRC's strategic goal to support U.S. interests in the safe and secure use of nuclear and radioactive materials and in nuclear nonproliferation.

Since the Fukushima accident, the NRC and its regulatory counterparts have shared a variety of information under the framework of these agreements, including preliminary results from the NRC's lessons learned activities. As the NRC's work in this area progresses and conclusions continue to develop, the NRC will continue to provide information about its activities and welcomes open, frequent exchanges of information to learn from its counterparts' efforts.

In the multilateral area, the U.S. has participated actively in nuclear safety activities at the IAEA and the NEA. This has included representation on nuclear safety committees and associated working groups, consultative work to develop international safety standards and guidance, and participation in international peer review missions. Following Fukushima, the U.S. has supported these agencies as they identify specific initiatives to address lessons learned and enhance multilateral communication, including identifying new or redistributing existing resources. In these areas, the U.S. has emphasized the importance of close cooperation among parties to maximize the effectiveness of the initiatives and avoid duplication of efforts.

6.4.2 Actions Taken

1. Following the event, the U.S. Government sent technical and diplomatic representatives to the U.S. Embassy in Japan to assist the U.S. Ambassador and his team. As requested, the U.S. representatives also assisted the Japanese Government with technical support.

2. The U.S. Government participated in major international meetings on Fukushima in 2011 including meetings hosted by the French Government as 2011 President of the Group of Eight Industrialized Nations (G8), and parallel meetings of the Group of Twenty (G20) in Paris, the NEA Forum in Paris, the IAEA ministerial-level conference in Vienna, and the European Nuclear Safety Regulators Group meeting in Brussels.

3. The U.S. Government participated in the development of the IAEA "Action Plan on Nuclear Safety," which was approved by the Board of Governors and adopted by the General Conference in September 2011.

4. The NRC is participating in an ongoing IAEA review of safety standards and guidance to address lessons learned from Fukushima.

5. The NRC participated in several IAEA international peer review missions to Japan, including the initial fact-finding mission that took place in May 2011.

6. The NRC, through its INES Advisory Committee member, is developing additional guidance on the application of INES in severe accidents as directed in the IAEA Action Plan on Nuclear Security.

7. The NRC is participating in meetings of the NEA Senior Task Group on the Impacts of Fukushima and in ongoing work within the NEA committee structure to determine NEA's role in post-Fukushima lessons-learned efforts.

8. The NRC hosted the October 2011 international workshop on lessons learned from Integrated Regulatory Review Service missions, which discussed the potential impacts of the Fukushima accident on the program, in keeping with the direction of the IAEA Action Plan on Nuclear Safety to enhance the international peer-review process.

9. The NRC held bilateral meetings with several regulatory counterparts to discuss lessons learned from Fukushima.

10. The NRC participated in the IAEA international experts meetings on reactor and spent fuel safety in light of the Fukushima accident in March 2012.

6.4.3 Actions Planned

1. Continue to participate in bilateral and multinational activities to exchange lessons learned from the Fukushima accident.

2. The U.S. will maintain regular contact with the Japanese regulator, the U.S. Embassy, and other stakeholders, as appropriate.

6.4.4 Schedule

Bilateral and multilateral activities for the exchange of lessons learned from Fukushima are ongoing. No schedule is necessary at this time beyond the current schedule of conferences, working group meetings, and consultancy meetings.

6.4.5 Results

U.S. participation in bilateral and multilateral activities such as those in the illustrative, but not comprehensive, list discussed above have provided fruitful opportunities for U.S technical and policy experts to exchange information and lessons learned with international counterparts who are employing a variety of approaches to address the enhancement of nuclear safety worldwide. As the global nuclear safety community continues to identify lessons from the Fukushima accident, these forums will continue to serve this beneficial purpose. These discussions have also enabled nuclear regulators, policymakers and other stakeholders to address the issue of how best to leverage resources to ensure that they are used efficiently while minimizing duplication and overlap.

6.5 Sharing Operating Experience

6.5.1 Discussion

As part of its safety goal, under the current NRC Strategic Plan, published in February 2012, the NRC is committed to the systematic evaluation of domestic and international operating experience and using information from those evaluations to improve NRC programs and for the benefit of U.S. commercial nuclear safety. As a result, the NRC's emphasis on the effective use and communication of international operating experience remains strong. It follows that, as part of its commitments to international stakeholders, and for the purposes of supporting international nuclear safety and security, the NRC shares U.S. nuclear industry operating experience with the international community through various means. The NRC participates in the INES and the International Reporting System for Operating Experience (IRS); both of these programs resulted from cooperation between the IAEA and the NEA of the OECD and make operating experience information accessible to the international nuclear community. NRC operating experience personnel also participate in a range of international meetings, conferences, and working groups, including the NEA/OECD/CNRA Working Group for Operating Experience (WGOE), the Advisory Committee for the IRS (the ACIRS), and workshops supporting IAEA and the European Commission's Operating Experience Clearinghouse.

NRC operating experience personnel rate all reactor related events using the criteria from the 2008 INES User's Manual. Furthermore, all nuclear or radiological events that are rated at an INES Level 2 or higher are transmitted to the IAEA for publication through its information channel on nuclear and radiological events, NEWS (http://www-news.iaea.org/). Additionally, the NRC posts all of the reactor-related INs to the IRS database in accordance with the most current manual for IRS coding. NRC INs are the vehicles the agency uses to communicate operating experience to U.S. commercial nuclear plant operators. Meetings, conferences and working groups also provide a multitude of opportunities for sharing information about specific U.S. events or trends that the NRC is following or to which the agency is responding. These interactions foster personal relationships between operating experience personnel from different countries and enable the sharing of operating experience through direct communications, such as e-mails and telephone conversations. Thus, the NRC operating experience staff routinely interacts with and responds to queries from international nuclear regulatory staff and associated bodies requiring assistance on operating experience topics. As an example, operating experience staff recently completed a comprehensive survey on operating experience organization, process, and products in response to an NEA-sponsored request from the regulatory authority in the United Arab Emirates.

After the reactor accidents at Fukushima Daiichi, the NRC shared relevant operating experience through various international forums. The NRC staff presented U.S. actions in response to Fukushima at various OECD/NEA/CNRA working group meetings, such as the WGOE and at meetings of the IRS. The NRC issued IN 2011-05, which describes the events at the Fukushima Daiichi plant with details that were known at the time. The IN also communicates several NRC rules that may be applicable from the perspective of the U.S. nuclear industry. The NRC shared this IN with the international community through the IRS. The NRC staff also developed a preliminary sequence of events that it shared within the NRC and with other U.S. Government agencies and entities that needed this information as the NRC sought to respond to and evaluate the event.

6.5.2 Actions Taken

1. IN 2011-05, "Tohoku-Taiheiyou-Oki Earthquake Effects on Japanese Nuclear Power Plants," dated March 18, 2011 was posted to the IRS database. IN 2011-05 describes the events at the Fukushima Daiichi Nuclear Power Station with details known at the time; the IN also communicates several NRC rules that maybe applicable from a U.S. nuclear industry perspective.

2. NRC operating experience personnel provided a presentation during the fall 2011 WGOE meeting that described the NRC's response to the events in Japan, the

radiological protective action recommendations made by the NRC to the U.S. Ambassador to Japan, the domestic considerations that were made in light of the events in Japan, and the NRC's NTTF findings.

3. In keeping with the NRC's value of openness, the agency has developed a Web site focused on communicating information related to the actions taken in response to the events in Japan. This site is easily accessible from the NRC's home page, www.nrc.gov.

6.5.3 Actions Planned

1. Continue supporting international nuclear safety and security through the open sharing of relevant and significant operating experience with international and domestic stakeholders.

6.5.4 Schedule

No schedule is necessary at this time, beyond the current schedule of conferences, working group meetings, and consultancy meetings. The NRC shares operating experience with the international community on a routine basis.

6.5.5 Results

While specific results concerning the sharing of operating experience are not readily measured or catalogued, it is accepted that the open and collaborative sharing of operating experience is always to the benefit of nuclear safety and security. From a domestic perspective, operating experience has served as a key component of the foundation upon which lessons learned from the nuclear events related to the Great East Japan Earthquake are developed. For example, insights from the extraordinary flooding conditions that occurred in the Midwestern U.S. in the spring of 2011 will be considered within the highest priority items of the NRC's response to the events in Japan. Similarly, insights gained from the sharing of operating experience among international partners has enabled fruitful exchanges and led to additional actions (e.g., consideration of the potential for the loss of UHS) not initially considered when prioritizing the NRC's actions in response to the events in Japan.

PART 2

Convention on Nuclear Safety 2012 Extraordinary Meeting:
U.S. Industry Response to the Fukushima Accident

May 2012

Institute of Nuclear Power Operations

INTRODUCTION

The U.S. industry response to the accident at the Fukushima Daiichi nuclear power plant (NPP) builds upon an existing regime of procedural guidance, equipment capability, and personnel training developed over many years. This regime helps maintain the integrity and performance of critical safety functions for bounding design-basis events and other beyond-design-basis events that may occur during the life of a plant. The following highlight several actions the U.S. industry has implemented over the last three decades to establish a robust capability to respond to these events:

1. The U.S. industry has adopted procedural guidance to address the preservation of critical safety features threatened by external events. This guidance includes the symptom-based emergency operating procedures (EOPs), developed in the 1980s following the Three Mile Island accident; severe accident management guidelines (SAMGs), developed in the 1990s following the Chernobyl accident; and the extensive damage management guidelines (EDMGs), developed in the 2000s following the terrorist attacks of September 11, 2001. This procedure regime provides defense-in-depth for design bases events and flexibility to respond to postulated conditions beyond the design basis of a plant.

2. The U.S. industry has addressed several specific issues related to external event response. These include procuring necessary equipment, developing procedural guidance, conducting training for plant operators to cope with a station blackout (SBO), managing and mitigating the effects of a terrorist attack (NRC Security Order B.5.b), and implementing emergency plans for responding to external events.

3. In accordance with NRC regulations, the U.S. industry implemented plant design modifications to increase AC power redundancy in the event of an SBO in the 1990s, and to establish robust containment hydrogen and pressure control capabilities. Additionally, the industry conducted individual plant examinations for external events (IPEEEs) using probabilistic risk assessment methods to identify external event vulnerabilities. It then put into place plant response strategies to mitigate those vulnerabilities.

With consideration to the above foundation, the response of the U.S. industry to the Fukushima nuclear accident involves both a response to orders and requests for information (RFIs) from the U.S. Nuclear Regulatory Commission (NRC) and the near-term, intermediate, and long-term actions industry initiatives as outlined below:

1. Prior to the NRC issuing orders and RFIs as a result of the lessons learned from Fukushima, a series of Institute of Nuclear Power Operations (INPO) event reports (IERs) defined the self-initiated near-term U.S. industry response. These include the following IERs, which are for internal use and available to INPO members only:

   • IER L1-11-1, "Fukushima Daiichi Nuclear Station Fuel Damage Caused by Earthquake and Tsunami," dated March 15, 2011

- IER L1-11-1, Supplement 1, "Fukushima Daiichi Nuclear Station Fuel Damage Caused by Earthquake and Tsunami," dated October 3, 2011

- IER L1-11-2 , "Fukushima Daiichi Nuclear Station Spent Fuel Pool Loss of Cooling and Makeup," dated April 25, 2011

- IER L1-11-3, "Weaknesses in Operator Fundamentals," dated June 15, 2011

- IER L1-11-4, "Near-Term Actions to Address the Effects of an Extended Loss of All AC Power in Response to the Fukushima Daiichi Event," dated August 1, 2011

2. A U.S. industry document, dated June 8, 2011, established the intermediate and long-term self-initiated industry response. Entitled "The Way Forward: U.S. Industry Leadership in Response to the Events at the Fukushima Daiichi Nuclear Power Plant," the document defines eight strategic goals for the U.S. industry in response to the Fukushima accident. This document is available to the public at http://nei.org/resourcesandstats/documentlibrary/safetyandsecurity/whitepaper/the-way-forward---june-2011.

The strategic goals are as follows:

1. The nuclear workforce remains focused on safety and operational excellence at all plants, particularly in light of the increased work that the response to the Fukushima event will represent.

2. Timelines for emergency response capability to ensure continued core cooling, containment integrity and spent fuel storage pool cooling to prevent fuel damage following SBO or challenges to the ultimate heat sink (UHS).

3. The U.S. nuclear industry is capable of responding effectively to any significant event in the U.S. with the response being scalable to support an international event, as appropriate.

4. SAMGs, security response strategies (B.5.b), and external event response plans are effectively integrated to ensure nuclear energy facilities are capable of a symptom-based response to events that could impact multiple reactors at a single site.

5. Margins for protection from external events are sufficient based on the latest hazards analyses and historical data.

6. Spent fuel pool (SFP) cooling and makeup functions are fully protective during periods of high heat load in the SFP and during extended SBO conditions.

7. Primary containment protective strategies can effectively manage and mitigate post-accident conditions, including elevated pressure and hydrogen concentrations.

8. Accident response procedures provide steps for controlling, monitoring and assessing potential radiation and ingestion pathways during and following an accident, including timely communication of accurate information.

In support of the strategic goals, "The Way Forward" defines six guiding principles and establishes an industry structure, including an executive steering committee, to implement the actions necessary to accomplish the strategic goals. The normal governance structure of the Electric Power Research Institute (EPRI), INPO, and the Nuclear Energy Institute (NEI) will oversee "The Way Forward" plan. The executive steering committee will use the strategic plan to assign actions and provide the appropriate accountability for action completion. The U.S. industry expects "The Way Forward" to be consistent with actions required as specified by the NRC's regulatory response.

TOPIC 1. EXTERNAL EVENTS

1.1 Overview of Actions Taken or Planned

In the area of external events, the U.S. industry undertook the following near-term actions in response to the Fukushima accident:

1. Based on the best information available at the time, on March 15, 2011, the U.S. industry issued IER L1-11-1, which contains the following recommendations:

    - Verify the capability to mitigate conditions that result from beyond-design-basis events, typically bounded by security threats, committed to as part of the NRC Security Order B.5.b issued on February 25, 2002 and in SAMGs.

    - Verify that the capability to mitigate SBO conditions required by station design is functional and valid.

    - Verify the capability to mitigate internal and external flooding events required by station design.

    - Perform walkdowns and inspections of important equipment needed to mitigate fire and flood events to identify the potential that the equipment's function could be lost during seismic events. Develop mitigating strategies for identified vulnerabilities.

    Actions for the above recommendations are to be in place within 30 days of the issuance of the IER.

2. Based on an industry review of discrepancies found during the completion of the walkdowns and inspections conducted for IER L1-11-1, the industry, on October 3, 2011, issued IER L1-11-1, Supplement 1. The supplement recommended actions to verify that equipment and mitigating strategies are included in station programs for configuration management, preventive maintenance, training, qualification, and routine audits and inspections. Actions to address identified issues will be provided to INPO in March 2012.

These near-term actions provided a high level of assurance that the U.S. industry was in conformance with plant design and appropriate industry initiatives and was prepared to respond to external events.

The U.S. industry has established the intermediate and long-term direction for external events with Strategic Goal 5 of "The Way Forward" plan. Strategic Goal 5 states: "Margins for protection from external events are sufficient based on the latest hazards analyses and historical data." The U.S. industry provides assurance that power and water requirements can be sustained to protect critical safety functions independent of a specific beyond-design-basis initiating event. The following ongoing actions support this strategic goal:

1. The chief nuclear officers in the U.S. industry have committed to validating external flooding and seismic design basis against the latest known data for external events using current methodology. This action is consistent with recommendations from the NRC Near-Term Task Force related to the accident at Fukushima. The validation includes the following:

    - Conduct evaluations and walkdowns to determine the capability to mitigate external flooding events, using approved procedures and acceptance criteria.

    - Conduct seismic evaluations and walkdowns on safety-related systems, structures, and components (SSCs), using approved procedures and acceptance criteria. Walkdowns will build on the walkdowns conducted for IER L1-11-1. The specific criteria and timing will be consistent with the anticipated regulatory requirements.

2. The consequences of postulated beyond-design-basis external events that are most impactful to reactor safety are loss of power and loss of the UHS. The U.S. industry is developing a strategy called "Diverse and Flexible Mitigation Capability" (FLEX) to address these events by providing multiple means of power and water supplies to support key safety functions. The FLEX strategy consists of portable equipment that provides the means of obtaining power and water to maintain or restore key safety functions for all reactors at a site and reasonable staging and protection of portable equipment from natural phenomena. FLEX uses equipment stored at the plant site and, if necessary, materials and equipment from either other plant sites or support centers for longer-term responses.

3. Based on the lessons learned from the Fukushima Daiichi accident, EPRI is assessing the risks and responses to a broad range of external hazards. This initiative includes a review of recent domestic events, such as flooding and earthquakes. Whereas the report's primary scope is on flooding and seismic events, it also considers the impact of high winds, transportation accidents, low temperatures, and loss of UHS.

1.2 Schedule of Planned Actions

1.2.1 Near-Term Actions

1. The U.S. industry completed actions specified in IER L1-11-1 in April 2011. NPPs documented any deficiencies or exceptions to the required actions in the site corrective action programs, to be resolved consistent with their safety priority. As of December 31, 2011, 95 percent of all deficiencies or exceptions had been corrected. The remaining items require specific plant conditions for resolution and will be addressed later in 2012 consistent with plant outage schedules.

1.2.2 Intermediate and Long-Term Actions

1. Consistent with the requirements of the NRC, the industry will conduct walkdowns and evaluate the adequacy of design features for mitigating external flooding.

2. Consistent with the requirements of the NRC, the industry will evaluate the adequacy and conduct walkdowns for seismic concerns.

3. Consistent with the requirements of the NRC, the industry will evaluate the adequacy and conduct walkdowns of safety-related SSCs for other external events, such as high winds, transportation accidents, low temperatures, and loss of UHS.

4. Under EPRI's leadership, the scope of work required for reviews of external hazards has been set, and working groups will start the actual reviews by September 2012.

1.3 Results of Completed Actions

1. By completing IER L1-11-1, the industry verified that equipment, procedures and staffing intended to implement beyond-design conditions are in place and functional for a single-unit external threat. The review of existing capabilities for design-basis and beyond-design-basis external events determined that surrounding area development, security modifications, and other site changes sometimes adversely affected external flood capability. Also, configuration management systems for procedures and drawings have not consistently captured beyond-design-basis commitments. The industry has corrected most specific deficiencies identified by IER L1-11-1, although a few are scheduled to be resolved later in 2012, consistent with plant outage schedules. Over 150 man-years of effort was expended during 2011 in conducting this work. The industry will address organizational lessons from the findings in an additional set of actions specified in IER L1-11-1, Supplement 1.

2. Industry responses to IER L1-11-1 Supplement 1 were provided to INPO in March 2012. These responses are not publicly available.

3. The industry has verified that individual station capabilities to mitigate an SBO event are in place and functional.

4. The industry has verified that individual station capabilities to mitigate design-basis internal and external flooding events are in place.

5. The industry has conducted walkdowns and inspections of equipment needed to mitigate floods and fires. NPPs have established mitigating strategies in the event of a flood or fire following a seismic event.

6. The industry self-identified, through inspections and reviews, portable equipment that improves capabilities to mitigate beyond-design-basis events. Equipment has either been acquired or was ordered by March 2012.

7. The industry has implemented enhanced operational capability for protecting spent fuel storage pools against extreme external events.

8. Consistent with the requirements of the NRC in Bulletin 2011-01, the industry provided information on maintaining, testing, and offsite support of mitigating strategies programs. The information confirmed compliance with requirements of the bulletin. Each plant's response to this bulletin is publicly available at http://www.nrc.gov/NRR/OVERSIGHT/ASSESS/mitigating-strategies.html.

TOPIC 2. DESIGN ISSUES

2.1 Overview of Actions Taken or Planned

The U.S. industry has initiated the following near-term actions to identify vulnerabilities with design and operating strategies involving the Fukushima Daiichi accident:

1. The U.S. industry issued IER L1-11-4, which recommends actions be taken to extend the time that existing equipment can be used to maintain critical safety functions for extended losses of AC power to allow more time for additional equipment to be provided to support long-term safe shutdown. Specific actions include the following:

   a. Develop methods to maintain or restore core cooling, containment integrity, and SFP inventory using existing installed and portable equipment during an extended loss of electrical AC power.

   b. Identify unit-specific information concerning coping response capability and design limitations for extended loss-of-power events; this information will be used in support of a broader U.S. response to the event.

   c. Identify essential instrumentation needed for monitoring core, containment integrity, and spent fuel safety; develop methods to ensure these functions are maintained during an extended loss of AC power.

   d. Develop methods to maintain or restore core cooling, containment, and SFP inventory using installed and portable equipment under extended loss of AC power conditions of at least 24 hours. The actual length of time critical safety functions can be maintained using existing installed and portable equipment will be analyzed and proposed upgrades and expected margin improvements will be reported. Key support functions -- such as instrumentation, fuel provisions, and maintenance -- are included in required actions.

2. Because there are many design variations among U.S. plants, the industry has established additional operational controls to enhance the design capability of SFPs. The reliability and availability of the SFP makeup systems have been analyzed specifically as part of IER L1-11-2. Short-term actions include increasing sensitivity to spent fuel storage event response and ensuring that NPPs maintain a high state of readiness to respond to events that challenge spent fuel storage integrity. Key actions taken include the following:

   a. Whenever the SFP time to reach 200 degrees Fahrenheit upon loss of normal cooling is less than 72 hours, identify and protect systems and equipment required to maintain decay heat removal and inventory control.

   b. Determine the time to reach 200 degrees Fahrenheit for all plant conditions in the event that normal cooling is lost, and maintain this information readily available to the control room and emergency response facilities.

c. Verify the adequacy of abnormal operating procedures (AOPs) to guide response to losses of all AC power and that the guidance can be implemented during severe weather, seismic events, loss of control room, and flood conditions.

d. Revise EOPs to provide precautions that SFP level and temperature should be monitored.

The U.S. industry has established the intermediate and long-term direction as follows:

Strategic Goals 2, 6, and 7 of "The Way Forward" establish the nuclear industry's response to design issues. These goals include associated principles and building blocks.

- Strategic Goal 2 states, "Timelines for emergency response capability to ensure continued core cooling, containment integrity and spent fuel storage pool cooling are synchronized to preclude fuel damage following station blackout."

- Strategic Goal 6 states, "Spent fuel pool cooling and makeup functions are fully protected during periods of high heat load in the spent fuel pool and during extended station blackout conditions."

- Strategic Goal 7 states, "Primary containment protective strategies can effectively manage and mitigate post-accident conditions, including elevated pressure and hydrogen concentrations."

The FLEX strategy adds another layer of safety to mitigate beyond-design-basis events. The objective of the strategy is to establish an essentially indefinite coping capability to prevent damage to the fuel in the reactor and SFP, and maintain the containment function by utilizing installed equipment, on-site portable equipment, and pre-staged off-site resources. This capability will address both extended losses of AC power and UHS events which could arise from beyond-design-basis external events. The diverse and flexible strategies focus on maintaining or restoring key safety functions and are not tied to any specific damage state or mechanistic assessment of external events.

The industry is reviewing procedures for existing boiling water reactor (BWR) Mark I containment vent valves, and it is evaluating the accessibility for operation of these valves. BWR Mark II plants will install hardened vent valves for containment. The industry will use the reviews to identify any needed improvements, which it will implement, consistent with NRC orders, plant operational schedules and its strategic plan. These actions will ensure that primary containment protective strategies can effectively manage and mitigate post-accident conditions, including elevated pressure and hydrogen concentrations. EPRI is supporting efforts related to understanding containment over-pressurization and combustible gas control.

In addition to the actions associated with IER L1-11-2, the industry is taking further actions to evaluate and identify the instrumentation and equipment needed to monitor spent fuel safety throughout an extended loss of AC power that includes depletion of DC power. NPPs will take appropriate site-specific actions to ensure continuity of cooling and makeup capability and remote temperature, level, and radiation monitoring of the SFP.

2.2 Schedule of Planned Actions

1. The U.S. industry developed plant-specific information concerning coping times and design limitations for extended loss-of-power events in January 2012. Consistent with the requirements of the NRC, it will develop procedural guidance for mitigating strategies for extended loss of AC and continuity of essential instrumentation and communication equipment.

2. Consistent with the requirements of the NRC, the U.S. nuclear industry will verify reliability of containment vents on a schedule that considers plant operating schedules. All actions are expected to be completed by December 31, 2016.

3. Consistent with the requirements of the NRC, the U.S. nuclear industry will evaluate and identify additional instrumentation and controls for SFPs. Any necessary plant modifications are expected to be completed by December 31, 2016.

4. Consistent with the requirements of the NRC, hardened vents will be in place for Mark I and Mark II containments by December 31, 2016.

2.3 Results of Completed Actions

1. The industry conducted plant-specific sensitivity studies for extended loss-of-power events as part of the IER 11-4 response, and provided results to INPO in January 2012.

2. The industry completed actions to increase sensitivity to spent fuel storage event response and ensure that NPPs maintain a high state of readiness to respond to events that challenge spent fuel storage integrity. These actions included the following:

    a. NPPs established controls to identify and protect systems and equipment required to maintain the functions of SFP decay heat removal and inventory control, such as clearly identifying protected equipment in the field and providing physical barriers whenever possible. This also included establishing controls and compensatory measures to prevent the SFP from reaching saturated conditions on loss of cooling when it is necessary to perform work on protected equipment.

    b. The industry established the time for the SFP to reach 200 degrees Fahrenheit (bulk temperature) in the event that normal cooling is lost and maintains this information in a format that is readily available in the control room and emergency response facilities.

    c. NPPs verified the adequacy of station AOPs for responding to the loss of SFP cooling and/or inventory and ensured these procedures include actions and contingencies to monitor SFP level and temperature and the capability to make up inventory to the SFPs during a loss of all AC power. This also includes verifying that NPPs can implement the guidance in the AOP during severe weather, seismic events, loss of control room, and flood conditions.

    d. NPPs revised station EOPs to include a precautionary statement that they should monitor SFP level and temperature.

TOPIC 3. SEVERE ACCIDENT MANAGEMENT AND RECOVERY (ONSITE)

3.1 Overview of Actions Taken or Planned

Strategic Goals 2 and 4 of "The Way Forward" establish the nuclear industry's response to severe accident management and recovery. These goals are as follows:

- Strategic Goal 2 states, "Timelines for emergency response capability to ensure continued core cooling, containment integrity, and spent fuel storage pool cooling are synchronized to preclude fuel damage following station blackout or challenges to the ultimate heat sink."

- Strategic Goal 4 states, "Severe accident management guidelines, security response strategies (B.5.b), and external event response plans are integrated effectively to ensure nuclear energy facilities are capable of a symptom-based response to events that could impact multiple reactors at a single site."

To address these strategic goals, the following actions have been taken or are under way:

1. Procedures - The industry has validated the ability of NPPs to implement EOPs, SAMGs, and B.5.b strategies. These actions included reviewing, identifying, and implementing enhancements to training programs to ensure that plant operators are able to implement the guidance. EPRI is leading an effort to update the technical bases of the SAMGs based on lessons learned from the Fukushima accident. The effort will determine whether the underlying assumptions and technical bases for the existing guidelines are adequate. The regime of EOPs, SAMGs, and EDMGs will be integrated as appropriate.

2. SFPs - As discussed in Section 2.1, actions associated with IER L1-11-2 increase sensitivity to spent fuel storage event response and ensure that NPPs maintain a high state of readiness to respond to events that challenge spent fuel storage integrity. Specific actions required are as follows:

    a. Verify the adequacy of AOPs to guide the response to the loss of all AC power and that the guidance can be implemented during severe weather, seismic events, loss of control room, and flood conditions.

    b. Revise EOPs to provide precautions that SFP level and temperature should be monitored.

3. AC Power - As described in Part II, Section 2 of this report, actions associated with IER L1-11-4 will extend the time that existing equipment can be used to maintain critical safety functions for extended losses of AC power until additional equipment can be provided to support long-term safe shutdown. Specific actions include the following:

    a. Sensitivity studies under the conditions specified in IER L1-11-4 will form the bases for plant-specific actions required to extend mitigating times with onsite equipment, as well as whether additional offsite equipment or material support are needed to extend mitigation times identified.

b. The industry will conduct further reviews for other outcomes (beyond extended loss of AC), such as concurrent loss of DC power or loss of UHS, to determine if the established mitigation strategy should be modified.

3.2 Schedule of Planned Actions

1. Consistent with the requirements of the NRC, the industry will update the technical bases of SAMGs. Based on the updated technical bases, the EOPs, EDMGs, and SAMGs will be integrated by 2016.

2. Consistent with the requirements of the NRC, site-specific actions in response to sensitivity studies will be taken to mitigate the effects of a complete loss of AC power. These actions are expected to be completed by December 31, 2016.

3. The industry will issue a supplement to IER L1-11-4, or other industry-approved guidance, once the industry results of the IER are analyzed and the necessity of further measures are identified. Further industry guidance will consider the FLEX strategy and align with NRC requirements. The need for and time of issuing this supplement is to be determined.

3.3 Results of Completed Actions

1. Actions associated with SFPs per IER L1-11-2 are complete. These include verifying the adequacy of AOPs to guide the response to a loss of all AC power and that the guidance can be implemented during severe weather, seismic events, loss of control room, and flood conditions. Also, the industry revised EOPs to provide precautions to monitor SFP level and temperature.

2. The U.S. industry has analyzed actions and set procedures for an extended loss of AC power, as defined by IER L1-11-4. Additionally, it has taken actions to ensure continuity of essential instrumentation and communication equipment under extended loss of AC power conditions.

3. The industry procured or ordered equipment needed for improving station response to or mitigation of beyond-design-basis events as of March 2012. The equipment ordered was identified as part of various reviews, including actions associated with IER L1-11-4. Equipment ordered includes portable pumps, generators, air compressors, switchgear and transformers, and other portable emergency equipment. Lists of equipment were provided to INPO in February 2012 (this information is not publicly available).

4. Consistent with the requirements of the NRC in Bulletin 2011-01, the industry verified plant capability for maintaining or restoring core cooling, containment, and spent fuel cooling under circumstances involving losses of large areas of the plants from explosions or fire. The information confirmed compliance with requirements of the bulletin. Each plant's response to this bulletin is publicly available at http://www.nrc.gov/NRR/OVERSIGHT/ASSESS/mitigating-strategies.html.

TOPIC 4. NATIONAL ORGANIZATIONS

4.1 Overview of Actions Taken or Planned

The roles and responsibilities for response to nuclear emergencies are described in "The United States Commercial Nuclear Industry Emergency Response Framework," which will be issued in March 2012. The Framework incorporates lessons learned from a self-assessment of the nuclear industry's response to the Fukushima Daiichi accident. Although the plan was written to respond to a domestic event, the U.S. nuclear industry used it in support of TEPCO during the Fukushima Daiichi accident. The Framework focuses on commercial and private organizations, but it also identifies Government agencies and their responsibilities as outlined in the Framework. Thirty recommendations were identified and broken down into immediate, near-term, and long-term actions.   The organizational roles and responsibilities of the industry are listed below:

1. Nuclear Facility Operator

    The owner/operator of a nuclear/radiological facility or materials has primary responsibility for mitigating the consequences of an incident; providing notification and appropriate protective action recommendations to State, local, and/or Tribal government officials; and minimizing the radiological hazard to the public.

    For incidents involving fixed facilities, the owner/operator has primary responsibility for actions within the facility boundary and may also have responsibilities for response and recovery activities outside the facility boundary under applicable legal obligations.

2. Electric Power Research Institute

    EPRI will maintain and provide independent technical expertise and event analysis capabilities. EPRI will maintain emergency response capabilities and be available for consultation and for conducting an in-depth analysis of the emergency.

3. Institute of Nuclear Power Operations

    INPO will serve as the industry notification point for industry events and will facilitate the US industry response to support the affected utility in mitigating the accident. INPO will maintain a dedicated emergency notification system for this purpose, as well as an emergency response center capable of supporting the functions of this plan. INPO will help identify and mobilize the resources of the nuclear industry in the event of an emergency, including obtaining and facilitating the flow of technical information from the affected utility to the nuclear industry and support organizations. INPO will provide operating experience to support the affected plant or the industry and liaisons to aid communication between the affected utility and stakeholder organizations.

4. Nuclear Energy Institute

    NEI will provide a broad industry perspective of the event to the Government, including Congress, the White House, and Federal agencies. NEI will supply professional communications and governmental relations support and resources to the affected site and support the management of media requests. NEI will maintain regular contact with Federal regulatory agencies and identify potential generic regulatory issues that may result from a significant plant event. NEI will coordinate responses to congressional or White House requests for information with the industry. NEI will cover the nature of and prognosis for the radiological situation causing the emergency and the actual or potential offsite radiological impact. Subsequent information should cover the status of mitigation, corrective actions, protective measures, and overall industry response to the emergency.

4.2 Schedule of Planned Actions

1. The "United States Commercial Nuclear Industry Emergency Response Framework" will be issued in March 2012 to address immediate lessons learned from the industry's response. Further revisions of this document will be issued to incorporate the near- and long-term lessons learned. A schedule for incorporating these lessons learned has yet to be determined.

4.3 Results of Completed Actions

1. A self-assessment of the industry response to the Fukushima accident was completed in December 2011. The self-assessment identified 14 principal lessons learned and 30 recommendations categorized as immediate, near-term, and long-term priorities. The results of the assessment are not publicly available.

TOPIC 5. EMERGENCY PREPAREDNESS AND RESPONSE AND POST-ACCIDENT MANAGEMENT (OFFSITE)

5.1 Overview of Actions Taken or Planned

Strategic Goals 3 and 8 of "The Way Forward" establish the nuclear industry's response to emergency preparedness (EP) and post-accident management. These goals include associated principles and building blocks.

- Strategic Goal 3 states, "The U.S. nuclear industry is capable of responding effectively to any significant event in the U.S. with the response being scalable to support an international event, as appropriate."

- Strategic Goal 8 states, "Accident response procedures provide steps for controlling, monitoring, and assessing potential radiation and ingestion pathways during and following an accident, including timely communication of accurate information."

To address these items, the following actions have been taken or are under way:

1. An industry initiative associated with "The Way Forward" improves the industry's ability to respond effectively to a significant event in the U.S., with a scalable response able to support an international event. There are two aspects to this initiative:

   a. The "United States Commercial Nuclear Industry Emergency Response Framework" was issued in March 2012 for industry senior executives to use when providing assistance and coordinating resources in the event of a severe nuclear emergency. The framework provides guidance and coordination and is fully integrated with the capabilities and responsibilities of the U.S. National Response Framework's Nuclear/Radiological Incident Annex. Utilities activate the plan in response to domestic or international nuclear incidents involving individual or multiple facilities, hostile actions toward a facility, or other incidents in which the full capabilities and assistance of the industry are needed to support the overall response. The guidance can be partially or fully implemented in the context of the event, in anticipation of a significant event, or in response to an event. Selective implementation allows for a graded response, delivery of the resources needed, and an appropriate level of coordination. This plan is described in greater detail as part of Topic 4.

   b. As part of the FLEX strategy, an initiative is under way to identify and store materials, supplies, and equipment needed to respond to a multi-unit site accident, placed in locations readily accessible to plants located in the U.S. This initiative is based on mitigating a long-term severe accident lasting beyond a plant's installed equipment's capabilities. The initiative includes information on sharing existing equipment and establishing the types of additional equipment (size and quantity) to be stored offsite. Onsite equipment is considered necessary for a plant to be self-sufficient in meeting its current design capabilities. Key aspects of the industry response capability are listed below:

      - staging equipment in central areas for regional response

- fitting equipment with standard mechanical and electrical connections
- sharing agreements between utilities
- providing adequate procedures and training

5.2 Schedule of Planned Actions

1. The initiative to identify and store materials, supplies, and equipment needed to respond to a multi-unit site accident, placed in regional centers that are readily accessible to plants located in the U.S., is under way. Contracts and protocols for regional centers for prestaged equipment are being developed consistent with the implementation of the FLEX strategy.

2. The industry plans to adopt the FLEX three-phased approach that addresses responding to extended loss of AC power and loss of UHS, using installed, onsite portable and offsite portable equipment. This strategy will be fully implemented by December 31, 2016.

5.3 Results of Completed Actions

1. The regional inventory for emergency equipment was established in February 2012. Standardized fittings for mechanical and electrical connections for use at any facility will be established in 2012. Portable equipment that improves capability to respond to beyond-design-basis events was identified as part of various reviews, including actions associated with IER L1-11-4. Lists of equipment were provided to INPO in February 2012, and includes portable pumps, generators, air compressors, switchgear and transformers, and other portable emergency equipment (this information is not publicly available). Many utilities have already acquired the identified equipment. The industry has ordered equipment not already procured for improving station response to or mitigation of beyond design basis events as of March 2012.

2. The "United States Commercial Nuclear Industry Emergency Response Framework" was issued in March 2012 to address lessons learned from the industry's response to the Fukushima event.

TOPIC 6. INTERNATIONAL COOPERATION

6.1 Overview of Actions Taken or Planned

The U.S. nuclear industry is fully committed to a high level of international engagement in its response to the Fukushima Daiichi nuclear accident. International engagement is inherent in the actions for all strategic goals in "The Way Forward" plan.

The following specific actions have been taken or are planned:

1. Share international approaches taken in response to the Fukushima Daiichi event. INPO and the World Association of Nuclear Operators (WANO) Atlanta Center co-hosted *The Fukushima Forum* on November 14-16, 2011. Twenty-three countries participated and presented a worldwide view of industry activities under way or planned. TEPCO representatives presented their findings of the accident at the opening of the forum. Participants also described specific aspects of their utility's responses to the accident, addressing actions associated with natural phenomena, SAMGs, losses of power, command and control, communications, emergency response, losses of UHS, SFPs, and hydrogen generation and containment. These findings informed participants of the activities under way at utilities around the world. Other similar forums will be held periodically.

2. The strong relationship between the Japanese and the U.S. nuclear industries at the time of the Fukushima accident was key to the high level of access to and support of TEPCO that the U.S. industry supplied from the early stages of the accident. The effectiveness of this level of support has been confirmed in many forums. The current levels of cooperation are based on the following programs:

    a. WANO, which consists of all NPP operators, provides a structure and forum for access to operating experience and direct interaction. The WANO Tokyo Center was key to the early and continuing information exchange between TEPCO and the rest of the world's nuclear operators and helped establish the interfaces necessary to bring U.S. industry support to Japan.

    b. INPO's International Participant Program includes a number of international nuclear power operating organizations that exchange information and experience, thereby promoting safe and reliable nuclear operations worldwide. This exchange was a substantial asset in supporting and learning from TEPCO because of the protocols in place at the time of the accident.

3. The U.S. industry, working closely with TEPCO, issued a sequence-of-events timeline describing the accident at Fukushima. The report, INPO 11-005, "Special Report on the Nuclear Accident at the Fukushima Daiichi Nuclear Power Station," is dated November 11, 2011. This timeline informs the U.S. and international response and is available to the public at http://nei.org/resourcesandstats/documentlibrary/safetyandsecurity/reports/special-report-on-the-nuclear-accident-at-the-fukushima-daiichi-nuclear-power-station.

4. The U.S. industry will continue to review the corrective actions of the international nuclear community to gain future insights from which the U.S. industry could benefit.

6.2 Schedule of Planned Actions

1. A second Fukushima Forum is scheduled for October 2012. Routine, weekly communications with foreign entities and counterparts will continue.

6.3 Results of Completed Actions

1. The Fukushima timeline, INPO 11-005, was issued in November 2011. The report provides a narrative overview and timeline for the earthquake, tsunami, and subsequent nuclear accident at Fukushima Daiichi on March 11, 2011. INPO 11-005 supplies an accurate, consolidated source of information on the sequence of events that occurred in the first days of the accident. The report reflects the best available information, most of which was obtained through direct and ongoing interaction with TEPCO. TEPCO, the Japanese Government, IAEA, and several other Japanese organizations provided the information used in the report. Data sources included logs, chart recorder indications, and personal accounts.

2. Results of the November 2011 Fukushima Forum, INPO 11-009, "Report of the INPO/WANO Fukushima Forum Proceedings," were published in December 2011. The document, which is for internal use and available to INPO members only, compiles industry issues and preventive and corrective actions associated with lessons learned from the Fukushima Daiichi accident. Topics include accident mitigation strategies, design issues, and emergency responses. Combined with meeting presentation materials, the report offers NPP operators an opportunity to benchmark their experience against industry peers and facilitate utility self-assessments.

REFERENCES

Code of Federal Regulations

Title 10 of the *Code of Federal Regulations* Part 50, "Domestic Licensing of Production and Utilization Facilities."

— — — Section 50.47, "Emergency Plans."

— — — Section 50.54, "Conditions of Licenses."

— — — Section 50.63, "Loss of All Alternating Current Power."

— — — Section 50.72, "Immediate Notification Requirements for Operating Nuclear Power Reactors."

— — — Appendix A, "General Design Criteria for Nuclear Power Plants."

— — — Appendix A, General Design Criteria 2, "Design Bases for Protection Against Natural Phenomena."

— — — Appendix E, "Emergency Planning and Preparedness for Production and Utilization Facilities."

— — — Appendix S, "Earthquake Engineering Criteria for Nuclear Power Plants."

Title 10 of the *Code of Federal Regulations* Part 52, "Licenses, Certifications, and Approvals for Nuclear Power Plants."

Title 10 of the *Code of Federal Regulations* Part 100, "Reactor Site Criteria."

— — — Section 20, "Factors To Be Considered When Evaluating Sites."

— — — Section 23, "Geologic and Seismic Siting Criteria."

— — — Subpart B, "Evaluation Factors for Stationary Power Reactor Site Applications on or After January 10, 1997."

— — — Appendix A, "Seismic and Geologic Siting Criteria for Nuclear Power Plants."

Institute of Nuclear Power Operations (INPO)

INPO 11-005, "Special Report on the Nuclear Accident at the Fukushima Daiichi Nuclear Power Station," November 11, 2011.

INPO 11-009, "Report of the INPO/WANO Fukushima Forum Proceedings," December 2011.

Institute of Nuclear Power Operations (INPO), Nuclear Energy Institute (NEI), and Electric Power Research Institute (EPRI)

"The Way Forward: U.S. Industry Leadership in Response to the Events at the Fukushima Daiichi Nuclear Power Plant," June 8, 2011.

International Atomic Energy Agency (IAEA)

"Action Plan on Nuclear Safety," September 2011.

"Convention on the Physical Protection of Nuclear Material, INFCIRC/274/Rev.1," May 1980.

"Convention on Early Notification of a Nuclear Accident, INFCIRC/335," November 1986.

"Convention on Assistance in the Case of a Nuclear Accident or Radiological Emergency, INFCIRC/336," November 1986.

"Convention on Nuclear Safety, INFCIRC/449," July 1994.

"Joint Convention on the Safety of Spent Fuel Management and on the Safety of Radioactive Waste Management, INFCIRC/546," December 1997.

U.S. Congress

Atomic Energy Act of 1954, as amended, 42 U.S.C. 2011 et seq.

National Environmental Policy Act of 1969, as amended, 42 U.S.C. 4321 et seq.

Nuclear Non-Proliferation Act of 1978, 22 U.S.C. 3201 et seq.

U.S. Department of Homeland Security

National Response Framework, January 2008.

U.S. Nuclear Regulatory Commission (NRC)

U.S. Nuclear Regulatory Commission, "Agreement Between the Government of the United States of America and the Government of Canada on Cooperation in Comprehensive Civil Emergency Planning and Management," and "Administrative Arrangement Between the United States Nuclear Regulatory Commission and the Atomic Energy Control Board of Canada for Cooperation and the Exchange of Information in Nuclear Regulatory Matters," June 21, 1989.

——— "Agreement for the Exchange of Information and Cooperation in Nuclear Safety Matters," and "Implementing Procedure for the Exchange of Technical Information and Cooperation in Nuclear Safety Matters Between the Nuclear Regulatory Commission of the United States of America and the Comisión Nacional de Seguridad Nuclear y Salvaguardias of Mexico," October 6, 1989.

— — — "Recommendations for Enhancing Reactor Safety in the 21st Century: The Near-Term Task Force Review of the Insights from the Fukushima Dai-Ichi Accident," July 12, 2011.

— — — Bulletin 2011-01, "Mitigating Strategies," May 11, 2011.

— — — Generic Issue 199, "Implications of Updated Probabilistic Seismic Hazard Estimates in Central and Eastern United States on Existing Plants," June 9, 2005.

— — — Draft Generic Letter 2011-XX, "Seismic Risk Evaluations for Operating Reactors," July 26, 2011 (Agencywide Documents Access and Management System Accession No. ML111710783).

— — — Generic Letter 87-02, "Verification of Seismic Adequacy of Mechanical and Electrical Equipment in Operating Reactors, Unresolved Safety Issue (USI) A-46," February 19, 1987.

— — — Generic Letter 88-20, Supplement 2, "Accident Management Strategies for Consideration in the Individual Plant Examination Process," April 4, 1990.

— — — Generic Letter 88-20, Supplement 4, "Individual Plant Examination of External Events (IPEEE) for Severe Accident Vulnerabilities," June 28, 1991.

— — — Generic Letter 89-16, "Installation of a Hardened Wetwell Vent," September 1, 1989.

— — — Information Notice 2010-018, "Generic Issue 199, 'Implications of Updated Probabilistic Seismic Hazard Estimates in Central and Eastern United States on Existing Plants,'" September 2, 2010.

— — — Information Notice 2011-05, "Tohoku-Taiheiyou-Oki Earthquake Effects On Japanese Nuclear Power Plants," March 18, 2011.

— — — Information Notice 2011-08, "Tohoku-Taiheiyou-Oki Earthquake Effects On Japanese Nuclear Power Plants – For Fuel Cycle Facilities," March 31, 2011.

— — — Interim Staff Guidance: "Compliance with 10 CFR 50.54(hh)(2) and 10 CFR 52.80(d) Loss of Large Areas of the Plant due to Explosions or Fires from a Beyond-Design Basis Event: DC/COL-ISG-016," June 9, 2010 (Official Use Only).

— — — Interim Staff Guidance: "Emergency Planning for Nuclear Power Plants," November 2011.

— — — NUREG-0654, "Criteria for Preparation and Evaluation of Radiological Emergency Response Plans and Preparedness in Support of Nuclear Power Plants," Rev. 1, November 1980.

— — — NUREG-0660, "NRC Action Plan Developed as a Result of the TMI-2 Accident," May 1980.

— — — NUREG-0737, "Clarification of TMI Action Plan Requirements," November 1980.

— — — NUREG-0737, Supplement 1, "Clarification of TMI Action Plan Requirements: Requirements for Emergency Response Capability," January 1983.

— — — NUREG-0800, "Standard Review Plan for the Review of Safety Analysis Reports for Nuclear Power Plants: LWR Edition, Rev. 6," March 2007.

— — — NUREG-1032, "Evaluation of Station Blackout Accidents at Nuclear Power Plants: Technical Findings Related to Unresolved Safety Issue A-44, Final Report," June 1988.

— — — NUREG-1614, Vol. 4, Strategic Plan: Fiscal Years 2008-2013," February 2008.

— — — NUREG-1650, "The United States of America Fifth National Report for the Convention on Nuclear Safety, Rev. 3," September 2010.

— — — NUREG/CR-6890, "Reevaluation of Station Blackout Risk at Nuclear Power Plants: Vol. 1, Analysis of Loss of Offsite Power Events: 1986-2004," December 2005.

— — — Regulatory Guide 1.29, "Seismic Design Classification," March 2007.

— — — Regulatory Guide 1.59, "Design Basis Floods for Nuclear Power Plants," August 1977.

— — — Regulatory Guide 1.60, "Design Response Spectra for Seismic Design of Nuclear Power Plants," December 1973.

— — — Regulatory Guide 1.102, "Flood Protection for Nuclear Power Plants," September 1976.

— — — Regulatory Guide 1.125, "Physical Models for Design and Operation of Hydraulic Structures and Systems for Nuclear Power Plants," March 2009.

— — — Regulatory Guide 1.155, "Station Blackout," August 1988.

— — — Regulatory Guide 1.208, "A Performance-Based Approach To Define the Site-Specific Earthquake Ground Motion," March 2007.

— — — Regulatory Guide 1.217, "Guidance for the Assessment of Beyond-Design-Basis Aircraft Impacts," August 2011.

— — — Security Order, "Issuance of Order for Interim Safeguards and Security Compensatory Measures for [Plant Name]," February 25, 2002.

— — — SECY-11-0053, "Final Rule: Enhancements to Emergency Preparedness Regulations (10 CFR Part 50 and 10 CFR Part 52)," April 8, 2011.

— — — SECY-11-0089, "Options for Proceeding with Future Level 3 Probabilistic Risk Assessment Activities," July 7, 2011.

— — — SECY-11-0093, "Near-Term Report and Recommendations for Agency Actions Following the Events in Japan" July 12, 2011

——— SECY 11-0117, "Proposed Charter for the Longer-Term Review of Lessons Learned from the March 11, 2011, Japanese Earthquake and Tsunami," August 26, 2011.

——— SECY-11-0124, "Recommended Actions to be taken Without Delay from the Near-Term Task Force Report," September 9, 2011

——— SECY-11-0137, "Prioritization of Recommended Actions to be Taken in Response to Fukushima Lessons Learned," October 3, 2011.

——— SECY-12-0025, "Proposed Orders and Requests for Information in Response to Lessons Learned from Japan's March 11, 2011, Great Tohoku Earthquake and Tsunami," February 17, 2012.

——— Tasking Memorandum COMGBJ-11-002, "NRC Actions Following the Events in Japan," March 23, 2011.

——— Temporary Instruction 2515/183, "Follow up to the Fukushima Daiichi Nuclear Station Fuel Damage Event," March 23, 2011.

——— Temporary Instruction 2515/184, "Availability and Readiness Inspection of Severe Accident Management Guidelines (SAMGs)," April 29, 2011.

U.S. Nuclear Regulatory Commission (NRC), U.S. Department of Energy (DOE), and Electric Power Research Institute (EPRI)

——— NUREG-2115, "Central and Eastern United States Seismic Source Characterization for Nuclear Facilities," January 2012.

Federal Register

74 FR 28112, "Consideration of Aircraft Impacts for New Nuclear Power Reactors; Final Rule," *Federal Register,* Vol. 74, Issue 112, pp. 28112-28147, Washington, DC, June 12, 2009.

Nuclear Energy Institute (NEI)

Nuclear Energy Institute, NEI 06-12, "B.5.b Phase 2 & 3 Submittal Guideline," Rev. 2, December 2006.

——— NEI 07-13, "Methodology for Performing Aircraft Impact Assessments for New Plant Designs," Rev. 7, May 2009.

APPENDIX A
SUMMARY TABLES

The tables included in the following pages provide a high-level summary of all activities taken or planned described in this report. The tables are divided by topic and are intended to assist in the development of the Coordinator's report.

TOPIC 1 – EXTERNAL EVENTS

Activity	Activities by the Operator			Activities by the Regulator		
	(Item 2.a) **Activity** - Taken? - Ongoing? - Planned?	(Item 2.b) **Schedule Or Milestones** for Planned Activities	(Item 2.c) **Results Available** - Yes? - No?	(Item 3.a) **Activity** - Taken? - Ongoing? - Planned?	(Item 3.b) **Schedule Or Milestones** for Planned Activities	(Item 3.c) **Conclusion Available** - Yes? - No?
Issued temporary instructions TI 2515/183 (responding to large area fires, explosions, SBO events, and flooding) and TI 2515/184 (SAMGs)	Taken. Industry self-initiated reviews of capabilities as part of IER 11-1	N/A	Yes. Responses to self-initiated reviews and inspections per TI2515/183 showed no significant findings.	Taken TI – issued March 2011	N/A	Yes, inspections completed. No significant findings
Issued Bulletin 2011-01 (mitigating strategies)	Taken Responses provided by July 2011	N/A	Yes. Licensees' responses provided. Information confirmed compliance with bulletin. (publicly available)	Taken. Bulletin – issued May 2011. Responses evaluated in March 2012 / Ongoing. Conclusion under development	Conclusion expected by summer 2012	No / No, conclusion under development
Prioritized NTTF recommendations				Taken	N/A	Yes, Action Tiers established
Engaged stakeholders to discuss seismic and flooding hazards				Taken	N/A	N/A

TOPIC 1 – EXTERNAL EVENTS

Activity	Activities by the Operator			Activities by the Regulator		
	(Item 2.a) Activity - Taken? - Ongoing? - Planned?	(Item 2.b) Schedule Or Milestones for Planned Activities	(Item 2.c) Results Available - Yes? - No?	(Item 3.a) Activity - Taken? - Ongoing? - Planned?	(Item 3.b) Schedule Or Milestones for Planned Activities	(Item 3.c) Conclusion Available - Yes? - No?
Requests for information regarding seismic and flooding hazards	Planned	Responses to RFIs to be provided consistent with NRC requirements and schedules	No. As a related activity, industry has verified that individual station capabilities to mitigate design basis flooding events are in place.	Taken – RFIs Issued March 12 / Planned	N/A / Implementation guidance for walkdowns – May 2012 / Guidance for hazards reevaluation – Nov 2012 / Subsequent orders and inspections may be required.	No / N/A / N/A / N/A
Engage stakeholders to discuss other external hazards				Planned	TBD	N/A
Requests for information regarding other external hazards	Planned	Responses to RFIs be provided consistent with NRC requirements and schedules	No	Planned	Issuing RFIs – TBD, 6mos after resources become available. Issuing orders and conducting inspections – TBD.	No

TOPIC 1 – EXTERNAL EVENTS

Activity	Activities by the Operator			Activities by the Regulator		
	(Item 2.a) **Activity?** - Taken? - Ongoing? - Planned?	(Item 2.b) **Schedule** **Or Milestones** for Planned Activities	(Item 2.c) **Results** **Available** - Yes? - No?	(Item 3.a) **Activity?** - Taken? - Ongoing? - Planned?	(Item 3.b) **Schedule** **Or Milestones** for Planned Activities	(Item 3.c) **Conclusion** **Available** - Yes? - No?
Perform multi-year Level 3 PRA				Ongoing	Plan under development. Implementation will take 4 years	No
Rulemaking on seismic and flooding hazards				Planned	TBD, schedule will be developed in July 2012	No
Enhancements related to seismically-induced fires and floods				Planned	TBD, schedule will be developed in July 2012	No
Take actions specified in IER L1-11-1, including verifying capability to mitigate conditions resulting from beyond design basis events, capability to mitigate SBO conditions, capability to mitigate internal and external flooding events, and walkdowns of equipment needed to mitigate fire and flooding events.	Taken Actions Completed April 2011	N/A	Yes, deficiencies documented in plant corrective action program (not publicly available)			
Take actions specified in IER 1-1 Supplement 1: Verify equipment and mitigating strategies are included in station programs for configuration management, preventive maintenance, training, qualification, and routine audits and inspections.	Taken		Responses provided to INPO in March 2012 (not publicly available)			
Working groups led by EPRI will conduct reviews of external hazards.	Planned	September 2012	No			

TOPIC 1 – EXTERNAL EVENTS

Activity	Activities by the Operator			Activities by the Regulator		
	(Item 2.a) **Activity** - Taken? - Ongoing? - Planned?	(Item 2.b) **Schedule** Or Milestones for Planned Activities	(Item 2.c) **Results** Available - Yes? - No?	(Item 3.a) **Activity** - Taken? - Ongoing? - Planned?	(Item 3.b) **Schedule** Or Milestones for Planned Activities	(Item 3.c) **Conclusion** Available - Yes? - No?
Acquire equipment for improving station response to or mitigation of beyond design basis events, as identified through IER 11-4 responses and other reviews.	Taken.		Lists of needed equipment were collated and provided to INPO in Feb 2011 (not publicly available) Identified equipment has either been acquired or ordered as of March 2012			

93

TOPIC 2 – DESIGN ISSUES

Activity	Activities by the Operator			Activities by the Regulator		
	(Item 2.a) Activity - Taken? - Ongoing? - Planned?	**(Item 2.b)** Schedule Or Milestones for Planned Activities	**(Item 2.c)** Results Available - Yes? - No?	**(Item 3.a)** Activity - Taken? - Ongoing? - Planned?	**(Item 3.b)** Schedule Or Milestones for Planned Activities	**(Item 3.c)** Conclusion Available - Yes? - No?
Issued temporary instruction TI 2515/183 (responding to large area fires, explosions, SBO events, and flooding)				Taken. TI – issued March 2011	N/A	Yes, inspections completed. No significant findings
Prioritized NTTF recommendations				Taken	N/A	Yes, action Tiers established
Rulemaking on SBO	Planned	Implementation of the rule will be scheduled consistent with NRC requirements.	N/A. The industry conducted plant-specific sensitivity studies for extended loss of power events as part of the IER 11-4 response, and provided results to INPO in Jan 2012. (not publicly available)	Taken - Advance notice of proposed rule issued in March 2012	N/A	No
	Ongoing	Meanwhile, industry is developing mitigating strategies for extended loss of AC and continuity of essential instrumentation and communication equipment. The issuance schedule is TBD		Planned	Rule to be completed in 2.5 yrs – Dec 2014	No

TOPIC 2 – DESIGN ISSUES

Activity	Activities by the Operator			Activities by the Regulator		
	(Item 2.a) **Activity?** - Taken? - Ongoing? - Planned?	(Item 2.b) **Schedule Or Milestones for Planned Activities**	(Item 2.c) **Results Available** - Yes? - No?	(Item 3.a) **Activity** - Taken? - Ongoing? - Planned?	(Item 3.b) **Schedule Or Milestones for Planned Activities**	(Item 3.c) **Conclusion Available** - Yes? - No?
Engage stakeholders and issue orders regarding SFP cooling	Planned	Comply with orders by Dec 2016	No	Taken – orders issued on March 12	N/A	No
				Planned	Orders full implementation - Dec 2016. Inspections – TBD.	No
Engage stakeholders and issue orders regarding hardened vents (Mark I and II)	Planned	Hardened vents for Mark I and Mark II containments to be installed by Dec 2016	No	Taken – orders issued on March 12	N/A	No
				Planned	Orders full implementation - Dec 2016. Inspections – TBD.	No
Engage stakeholders to discuss regulatory actions on filtration of containment vents	Planned	Responses to be provided consistent with NRC requirements and schedules	No	Planned	SECY paper will be issued in July 2012	No
Evaluate need for reliable vents in other containments (Not Mark I and II)				Planned	TBD	No

TOPIC 2 – DESIGN ISSUES

Activity	Activities by the Operator			Activities by the Regulator		
	(Item 2.a) Activity - Taken? - Ongoing? - Planned?	(Item 2.b) Schedule Or Milestones for Planned Activities	(Item 2.c) Results Available - Yes? - No?	(Item 3.a) Activity - Taken? - Ongoing? - Planned?	(Item 3.b) Schedule Or Milestones for Planned Activities	(Item 3.c) Conclusion Available - Yes? - No?
Evaluate need for enhanced hydrogen control and mitigation				Planned	TBD	No
Engage stakeholders and issue orders regarding reliable SFP instrumentation	Planned	Additional SFP instrumentation to be installed as needed by Dec 2016 in accordance with NRC orders	No	Taken – orders issued on March 12	N/A	No
				Planned	Orders full implementation - Dec 2016. Inspections – TBD.	No
Rulemaking on SFP makeup capability and water level instrumentation				Planned	5 yrs	No
Engage stakeholders to discuss UHS assumptions				Taken	N/A	N/A
Issue requests for information on seismic and flooding hazards, including UHS systems	Planned	Responses to be provided consistent with NRC requirements and schedules	No	Taken – RFIs issued March 12	N/A	No
				Planned	Subsequent orders and inspections may be required.	N/A

TOPIC 2 – DESIGN ISSUES

Activity	Activities by the Operator			Activities by the Regulator		
	(Item 2.a) Activity - Taken? - Ongoing? - Planned?	(Item 2.b) Schedule Or Milestones for Planned Activities	(Item 2.c) Results Available - Yes? - No?	(Item 3.a) Activity - Taken? - Ongoing? - Planned?	(Item 3.b) Schedule Or Milestones for Planned Activities	(Item 3.c) Conclusion Available - Yes? - No?
Orders regarding beyond-design-basis events, including loss of UHS	Planned	Comply with orders by Dec 2016	No	Taken – orders issued on March 12 / Planned	N/A / Orders full implementation - Dec 2016. Inspections – TBD.	No / No
Conduct sensitivity studies to determine capabilities for responding to extended loss of power events.	Taken. Completed in Jan 2012	N/A	Yes. (not publicly available)			
Increase sensitivity to spent fuel storage event response and ensure that a high state of readiness is maintained to events.	Taken. Completed in Nov 2011	N/A	Controls established to identify/ protect systems/equipment required to maintain the functions of SFP decay heat removal and inventory control. Procedures revised where needed.			

TOPIC 3 – SEVERE ACCIDENT MANAGEMENT AND RECOVERY (ON-SITE)

Activity	Activities by the Operator			Activities by the Regulator		
	(Item 2.a) Activity - Taken? - Ongoing? - Planned?	(Item 2.b) Schedule Or Milestones for Planned Activities	(Item 2.c) Results Available - Yes? - No?	(Item 3.a) Activity - Taken? - Ongoing? - Planned?	(Item 3.b) Schedule Or Milestones for Planned Activities	(Item 3.c) Conclusion Available - Yes? - No?
Prioritized NTTF recommendations				Taken	N/A	Yes, action Tiers established
Training to NRC staff and Resident Inspectors on SAMGs				Planned	Schedule will be issued in 1 yr	No
Issued information notice on earthquake and tsunami				Taken. IN – issued March 2011	N/A	N/A
Issued temporary instruction TI 2515/184 (SAMGs)				Taken. TI – issues April 2011	N/A	Yes, inspections completed. No significant findings

TOPIC 3 – SEVERE ACCIDENT MANAGEMENT AND RECOVERY (ON-SITE)

Activity	Activities by the Operator			Activities by the Regulator		
	(Item 2.a) **Activity** - Taken? - Ongoing? - Planned?	(Item 2.b) **Schedule** Or Milestones for Planned Activities	(Item 2.c) **Results Available** - Yes? - No?	(Item 3.a) **Activity** - Taken? - Ongoing? - Planned?	(Item 3.b) **Schedule** Or Milestones for Planned Activities	(Item 3.c) **Conclusion Available** - Yes? - No?
Issued Bulletin 2011-01 (mitigating strategies)	Taken Responses provided by July 2011	N/A	Yes. Licensees' responses provided. Information confirmed compliance with bulletin. (publicly available)	Taken. Bulletin – issued May 2011. Responses evaluated in March 2012	N/A	No
				Ongoing. Conclusion under development	Conclusion expected by summer 2012	No, conclusion under development
				Planned. Use insights to develop guidance for orders on mitigating strategies	Aug 2012	No, guidance under development

TOPIC 3 – SEVERE ACCIDENT MANAGEMENT AND RECOVERY (ON-SITE)

Activity	Activities by the Operator			Activities by the Regulator		
	(Item 2.a) Activity - Taken? - Ongoing? - Planned?	(Item 2.b) Schedule Or Milestones for Planned Activities	(Item 2.c) Results Available - Yes? - No?	(Item 3.a) Activity - Taken? - Ongoing? - Planned?	(Item 3.b) Schedule Or Milestones for Planned Activities	(Item 3.c) Conclusion Available - Yes? - No?
Rulemaking on licensee emergency response to coordinate EOPs, SAMGs, EDMGs	Planned	Implementation of the rule will be scheduled consistent with NRC requirements.	No	Planned	5 yrs	No
	Industry to voluntarily integrate EOPs, EDMGs, and SAMGs	2016	No			
Expand regulatory framework to include beyond-design-basis events and make changes to ROP				Planned	SECY Paper - Feb 2013. ROP changes – TBD.	No
Engage stakeholders and issue Orders related to: - beyond design basis events and multi-unit events - hardened vents for Mark I and II - SFP instrumentation	Planned	Compliance with orders by Dec 2016	No	Planned	Orders full implementation - Dec 2016. Inspections – TBD.	No

TOPIC 3 – SEVERE ACCIDENT MANAGEMENT AND RECOVERY (ON-SITE)

Activity	Activities by the Operator			Activities by the Regulator		
	(Item 2.a) Activity - Taken? - Ongoing? - Planned?	(Item 2.b) Schedule Or Milestones for Planned Activities	(Item 2.c) Results Available - Yes? - No?	(Item 3.a) Activity - Taken? - Ongoing? - Planned?	(Item 3.b) Schedule Or Milestones for Planned Activities	(Item 3.c) Conclusion Available - Yes? - No?
Rulemaking on prolonged SBOs	Planned Improve plant capability to mitigate effects of a complete loss of AC power	Implementation of rule to be scheduled consistent with NRC requirements and schedules	No	Planned	5 yrs	No
Evaluate regulatory action for - hardened vents for other containments (not Mark I and II) - hydrogen control and mitigation				Planned	Schedule TBD in summer 2012	No
The technical bases of severe accident management guidelines will be updated.	Planned	To be scheduled consistent with NRC requirements	No			
Evaluate the need for supplement to IER L1-11-4 *Near-Term Actions to Address the Effects of an Extended Loss of All AC Power in Response to the Fukushima Daiichi Event,*	Planned Utility responses to IER L1-11-4 will be used to determine if a supplement is necessary.	TBD	No			

101

TOPIC 3 – SEVERE ACCIDENT MANAGEMENT AND RECOVERY (ON-SITE)

Activity	Activities by the Operator			Activities by the Regulator		
	(Item 2.a) Activity - Taken? - Ongoing? - Planned?	(Item 2.b) Schedule Or Milestones for Planned Activities	(Item 2.c) Results Available - Yes? - No?	(Item 3.a) Activity - Taken? - Ongoing? - Planned?	(Item 3.b) Schedule Or Milestones for Planned Activities	(Item 3.c) Conclusion Available - Yes? - No?
Complete the actions of IER L1-11-2, *Fukushima Daiichi Nuclear Station Spent Fuel Pool Loss of Cooling and Makeup*	Taken Completed by Nov 2011	N/A	Yes. Results of actions taken were provided to INPO. Sites revised procedures to strengthen defences of SFP temperatures and levels, addressed training of personnel, and have taken steps to monitor level and temperature during extended SBO.			
Complete the actions of IER L1-11-4, *Near-Term Actions to Address the Effects of an Extended Loss of All AC Power in Response to the Fukushima Daiichi Event,*	Ongoing Industry responses were received in Jan 2012.	Actions will be determined based on analysis of responses.	No			

TOPIC 3 – SEVERE ACCIDENT MANAGEMENT AND RECOVERY (ON-SITE)

Activity	Activities by the Operator			Activities by the Regulator		
	(Item 2.a) Activity - Taken? - Ongoing? - Planned?	(Item 2.b) Schedule Or Milestones for Planned Activities	(Item 2.c) Results Available - Yes? - No?	(Item 3.a) Activity - Taken? - Ongoing? - Planned?	(Item 3.b) Schedule Or Milestones for Planned Activities	(Item 3.c) Conclusion Available - Yes? - No?
Acquire equipment for improving station response to or mitigation of beyond design basis events, as identified through IER 11-4 responses and other reviews.	Ongoing	Not already procured equipment was ordered by March 2012	Yes. Lists of needed equipment were collated and provided to INPO in Feb 2012 (not publicly available) Over 450 major pieces of portable equipment has been procured or ordered, including diesel-driven pumps, diesel-driven generators, trucks or trailers, switchgear, air compressors, and other gear.			

TOPIC 4 – NATIONAL ORGANIZATIONS

Activity	Activities by the Operator			Activities by the Regulator		
	(Item 2.a) Activity - Taken? - Ongoing? - Planned?	(Item 2.b) Schedule Or Milestones for Planned Activities	(Item 2.c) Results Available - Yes? - No?	(Item 3.a) Activity - Taken? - Ongoing? - Planned?	(Item 3.b) Schedule Or Milestones for Planned Activities	(Item 3.c) Conclusion Available - Yes? - No?
Issued Information Notice on earthquake and tsunami				Taken IN – issued March 2011	N/A	N/A
Issued temporary instruction 2515/183 (responding to large area fires, explosions, SBO events, and flooding) and TI 2515/184 (SAMGs)				Taken TIs issued March and April 2011	N/A	Yes, inspections completed. No significant findings
Issued Bulletin 2011-01 (mitigating strategies)	Taken Responses provided by July 2011	N/A	Yes. Licensees' responses provided. Information confirmed compliance with bulletin. (publicly available)	Taken. Bulletin – issued May 2011. Responses evaluated in March 2012	N/A	No
				Ongoing. Conclusion under development	Conclusion expected by summer 2012	No, conclusion under development
				Planned. Use insights to develop guidance for orders on mitigating strategies	Aug 2012	No, guidance under development

TOPIC 4 – NATIONAL ORGANIZATIONS

Activity	Activities by the Operator			Activities by the Regulator		
	(Item 2.a) Activity - Taken? - Ongoing? - Planned?	(Item 2.b) Schedule Or Milestones for Planned Activities	(Item 2.c) Results Available - Yes? - No?	(Item 3.a) Activity - Taken? - Ongoing? - Planned?	(Item 3.b) Schedule Or Milestones for Planned Activities	(Item 3.c) Conclusion Available - Yes? - No?
Issue temporary instruction 2600/0015 (fuel facilities)				Taken – TI issued Sept 2011	N/A	No
				Planned	Inspections to be completed by Sept 2012	No
Established the NRC Task Force and developed recommendations				Taken	N/A	Yes, NTTF report available
Engaged stakeholders to discuss the Task Force recommendations and prioritized recommendations				Taken	N/A	Yes, Action Tiers established
Engage stakeholders, and issue orders and requests for information for near-term activities				Taken – orders and RFIs issued March 12	N/A	No
Established the Japan Lessons Learned Directorate				Taken	N/A	N/A
Support Members of the Congress and testify in hearings				Taken Ongoing	N/A, support provided as needed	N/A
Held over 10 public meetings to discuss NRC actions and Task Force recommendations				Taken Ongoing	Additional meetings will be scheduled as needed	N/A

TOPIC 4 – NATIONAL ORGANIZATIONS

Activity	Activities by the Operator			Activities by the Regulator		
	(Item 2.a) **Activity** - Taken? - Ongoing? - Planned?	(Item 2.b) **Schedule** Or Milestones for Planned Activities	(Item 2.c) **Results Available** - Yes? - No?	(Item 3.a) **Activity** - Taken? - Ongoing? - Planned?	(Item 3.b) **Schedule** Or Milestones for Planned Activities	(Item 3.c) **Conclusion Available** - Yes? - No?
Issue *The United States Commercial Nuclear Industry Emergency Response Framework* to address immediate lessons learned from the industry's response.	Taken – March 2012	N/A	N/A	Planned The government is also updating its response framework document "National Response Framework" (Discussed in Part 1, Topic 5)	2012	No
Conduct self-assessment of the industry response to the Fukushima accident	Complete	N/A	Yes. Actions categorized as immediate, near-term, and long-term. (not publicly available)			

TOPIC 5 – EMERGENCY PREPAREDNESS

Activity	Activities by the Operator			Activities by the Regulator		
	(Item 2.a) Activity - Taken? - Ongoing? - Planned?	(Item 2.b) Schedule Or Milestones for Planned Activities	(Item 2.c) Results Available - Yes? - No?	(Item 3.a) Activity - Taken? - Ongoing? - Planned?	(Item 3.b) Schedule Or Milestones for Planned Activities	(Item 3.c) Conclusion Available - Yes? - No?
Prioritized NTTF recommendations				Taken	N/A	Yes, action Tiers established
Engaged stakeholders and issue requests for information related to: - staffing requirements during an emergency - communications equipment	Taken		Identified equipment for improving station response to or mitigation of beyond design basis events, as identified through IER 11-4 responses and other reviews was acquired or ordered by March 2012	Taken – RFIs Issued March 12	N/A	No
				Planned	Subsequent orders and inspections may be required.	N/A
Evaluate rulemaking related to training and exercises for multi-units and prolonged SBOs				Planned	TBD	No
Complete ERDS Virtual Private Network transition				Planned	June 2012	No
Evaluate regulatory options to address monitoring capability for ERDS information				Planned	TBD	No

107

TOPIC 5 – EMERGENCY PREPAREDNESS

Activity	Activities by the Operator			Activities by the Regulator		
	(Item 2.a) Activity - Taken? - Ongoing? - Planned?	(Item 2.b) Schedule Or Milestones for Planned Activities	(Item 2.c) Results Available - Yes? - No?	(Item 3.a) Activity - Taken? - Ongoing? - Planned?	(Item 3.b) Schedule Or Milestones for Planned Activities	(Item 3.c) Conclusion Available - Yes? - No?
Evaluate regulatory options to address multi-unit dose assessment				Planned	Schedule will be developed by Summer 2012	No
Continue studies on health effects on populations around NPPs				Planned	Schedule will be developed by Summer 2012	No
Updated crisis communication plan and developed a standard operating protocol				Taken	N/A	Yes
Conduct post-event After Action Review (AAR)				Taken – AAR conducted / Planned	N/A / Action items being prioritized. Implementation schedule TBD	No
Develop Public Inquiry Desk roster	Ongoing	March 2012	No	Planned	TBD	No

TOPIC 5 – EMERGENCY PREPAREDNESS

Activity	Activities by the Operator			Activities by the Regulator		
	(Item 2.a) **Activity** - Taken? - Ongoing? - Planned?	(Item 2.b) **Schedule** Or Milestones for Planned Activities	(Item 2.c) **Results** Available - Yes? - No?	(Item 3.a) **Activity** - Taken? - Ongoing? - Planned?	(Item 3.b) **Schedule** Or Milestones for Planned Activities	(Item 3.c) **Conclusion** Available - Yes? - No?
Update the U.S. Homeland Security's National Response Framework	Taken – March 2012 The industry updated its response framework document "*The United States Commercial Nuclear Industry Emergency Response Framework*" (Discussed in Part 2, Topic 4)		N/A	Planned	2012	No
Evaluating regulatory options related to: - protective equipment for emergency responders - command and control structure - evaluate ERDS and determine alternate methods for transmittal - enhancements to site-recovery and site-entry plans - study of efficacy of real-time - conduct training on radiation safety and use of KI around the NPPs				Planned	TBD, schedule will be issued in July 2012	No

TOPIC 5 – EMERGENCY PREPAREDNESS

Activity	Activities by the Operator			Activities by the Regulator		
	(Item 2.a) Activity - Taken? - Ongoing? - Planned?	(Item 2.b) Schedule Or Milestones for Planned Activities	(Item 2.c) Results Available - Yes? - No?	(Item 3.a) Activity - Taken? - Ongoing? - Planned?	(Item 3.b) Schedule Or Milestones for Planned Activities	(Item 3.c) Conclusion Available - Yes? - No?
Identify and store materials, supplies, and equipment needed to respond to a multi-unit site accident in locations that are readily accessible to plants located in the U.S is underway.	Ongoing	Contracts and protocols for regional centers for prestaged equipment are being developed in conjunction with FLEX strategy.	No. Materials and costs have been identified. Regional inventory for emergency equipment established in Feb 2012.			
Adopt the three-phased approach of the diverse flexible plan that addresses improving responses during an extended loss of AC power and loss of UHS, using installed, on-site portable, and off-site portable equipment.	Ongoing	To be scheduled consistent with the NRC requirements	No			
Establish standardized fittings for mechanical and electrical connections to allow use at any facility.	Taken	N/A	N/A			

TOPIC 6 – INTERNATIONAL COOPERATION

Activity	Activities by the Operator			Activities by the Regulator		
	(Item 2.a) Activity - Taken? - Ongoing? - Planned?	(Item 2.b) Schedule Or Milestones for Planned Activities	(Item 2.c) Results Available - Yes? - No?	(Item 3.a) Activity - Taken? - Ongoing? - Planned?	(Item 3.b) Schedule Or Milestones for Planned Activities	(Item 3.c) Conclusion Available - Yes? - No?
Maintain CNS leadership roles (Vice President, Vice Chair Group 1) for Extraordinary Meeting				Taken Ongoing	N/A	N/A
Participate in discussions related to changes to conventions				Ongoing	N/A	N/A
Participate in IAEA Action Plan on Nuclear Safety				Taken	N/A	N/A
Appoint National Competent Authority contact for the Early Convention and Assistant Convention				Taken	N/A	N/A
Support Preparatory Committee meeting of Parties to the Non-Proliferation Treaty				Ongoing	N/A	N/A
Support ratification of the amendment to the Convention on Physical Protection of Nuclear Material				Ongoing	N/A	N/A
Submit CNS proposals				Taken	April 2012	N/A
Participate in - Joint Convention - CNS Extraordinary Meeting - CNS 6th Review Meeting - Non-Proliferation Treaty Preparatory Committee Meeting and Review Conference				Planned	JC –May 2012 EM –Aug 2012 6th CNS- 2014 NPT – 2013-2015	N/A

TOPIC 6 – INTERNATIONAL COOPERATION

Activity	Activities by the Operator			Activities by the Regulator		
	(Item 2.a) Activity - Taken? - Ongoing? - Planned?	(Item 2.b) Schedule Or Milestones for Planned Activities	(Item 2.c) Results Available - Yes? - No?	(Item 3.a) Activity - Taken? - Ongoing? - Planned?	(Item 3.b) Schedule Or Milestones for Planned Activities	(Item 3.c) Conclusion Available - Yes? - No?
Exchange information with Canada and Mexico				Taken- Exchanged information. Bilateral with Canada and Mexico completed. Planned Ongoing	N/A Bilateral with Canada and Mexico –2012 Support to be provided, as needed.	N/A N/A N/A
Support to U.S. Embassy in Japan				Taken Ongoing	Support to be provided, as needed.	N/A
Participated in - G8/G20 ministerial meeting - NEA forum - IAEA ministerial conference - ENSREG meeting - IAEA review of safety standards - IAEA Japan fact-finding mission - IAEA International Experts meetings				Taken – Multiple dates Ongoing	Support to be provided, as needed. NRC will continue to attend CNRA STG meetings	N/A
Develop guidance on application of INES				Ongoing	TBD	N/A

112

	TOPIC 6 – INTERNATIONAL COOPERATION					
	Activities by the Operator			Activities by the Regulator		
	(Item 2.a)	(Item 2.b)	(Item 2.c)	(Item 3.a)	(Item 3.b)	(Item 3.c)
Activity	Activity - Taken? - Ongoing? - Planned?	Schedule Or Milestones for Planned Activities	Results Available - Yes? - No?	Activity - Taken? - Ongoing? - Planned?	Schedule Or Milestones for Planned Activities	Conclusion Available - Yes? - No?
Hosted IRRS lessons learned workshop and discussed impacts of Fukushima				Taken - October 2011	N/A	N/A
Support bilateral meetings on lessons learned from Fukushima				Taken – Multiple dates / Ongoing	Will continue to provide support, as needed	N/A
Issued IN 2011-05 (earthquake and tsunami)				Taken. IN – issued March 2011	N/A	N/A
Participated in WGOE meeting - presented the NRC response to Fukushima				Taken - Fall 2011	N/A	N/A
Developed NRC public website on actions taken in response to Fukushima				Taken	N/A	N/A
Hold forums for international participants to share lessons and actions taken in response to Fukushima accident.	Taken - First forum Nov 2011 / Planned	2nd forum – Oct 2012	Yes. Specific utility responses discussed. Key aspects shared in a private document – Dec 2012.			
Issue Fukushima accident timeline.	Taken	N/A	Yes. Report publicly available – Nov 2011			

NRC FORM 335	U.S. NUCLEAR REGULATORY COMMISSION	1. REPORT NUMBER
(12-2010) NRCMD 3.7 **BIBLIOGRAPHIC DATA SHEET** *(See instructions on the reverse)*		(Assigned by NRC, Add Vol., Supp., Rev., and Addendum Numbers, if any.) NUREG-1650, Rev 4

2. TITLE AND SUBTITLE

The United States of America National Report for the 2012 Convention on Nuclear Safety Extraordinary Meeting

3. DATE REPORT PUBLISHED

MONTH	YEAR
July	2012

4. FIN OR GRANT NUMBER

5. AUTHOR(S)

U.S. Nuclear Regulatory Commission (NRC)
Institute of Nuclear Power Operations (INPO)

6. TYPE OF REPORT

Technical

7. PERIOD COVERED (Inclusive Dates)

March 2011 - May 2012

8. PERFORMING ORGANIZATION - NAME AND ADDRESS (If NRC, provide Division, Office or Region, U. S. Nuclear Regulatory Commission, and mailing address; if contractor, provide name and mailing address.)

Division of Inspections and Regional Support
Office of Nuclear Reactor Regulations
U.S. Nuclear Regulatory Commission
Washington, D.C. 20555-0001

9. SPONSORING ORGANIZATION - NAME AND ADDRESS (If NRC, type "Same as above", if contractor, provide NRC Division, Office or Region, U. S. Nuclear Regulatory Commission, and mailing address.)

Same as above (item 8)

10. SUPPLEMENTARY NOTES

11. ABSTRACT (200 words or less)

The United States (U.S.) Nuclear Regulatory Commission (NRC), in coordination with the U.S. Department of State, U.S. Department of Energy, and the Institute of Nuclear Power Operations, prepared this report, "The United States of America National Report for the 2012 Convention on Nuclear Safety Extraordinary Meeting," and will submit it for peer review at the 2012 Convention on Nuclear Safety (CNS) Extraordinary Meeting to be held at the International Atomic Energy Agency in Vienna, Austria, in August 2012. This report addresses the actions taken by the U.S. to improve nuclear safety in response to the March 11, 2011, accident at the Fukushima Daiichi nuclear power plant in Japan. It demonstrates how the U.S. contributes to achieving and maintaining a high level of nuclear safety worldwide by enhancing national measures and international cooperation and by meeting the obligations of all the articles established by CNS. It describes how the U.S. addressed six topics in relation to the Fukushima accident: (1) external events, (2) design issues, (3) severe accident management and recovery, (4) national organizations, (5) emergency preparedness and response and post-accident management, and (6) international cooperation. Similar to the U.S. National Report for the fifth CNS issued in 2010, this report includes a section developed by the Institute of Nuclear Power Operations describing work done by the U.S. nuclear industry in response to the Fukushima accident.

12. KEY WORDS/DESCRIPTORS (List words or phrases that will assist researchers in locating the report.)

KEY WORDS:Fukushima, Seismic, Flooding, Preventing Loss of Onsite, Power, Japan, Earthquake, Tsunami, Multi-Unit Events, International, Design Issues, Station Blackouts, Equipment Availability, Overpressure, Containment, and Protection

13. AVAILABILITY STATEMENT

unlimited

14. SECURITY CLASSIFICATION

(This Page)

unclassified

(This Report)

unclassified

15. NUMBER OF PAGES

16. PRICE

UNITED STATES
NUCLEAR REGULATORY COMMISSION
WASHINGTON, DC 20555-0001

—————

OFFICIAL BUSINESS

NUREG-1650
Revision 4

The United States of America National Report for the 2012 Convention on Nuclear Safety Extraordinary Meeting

July 2012

www.ingramcontent.com/pod-product-compliance
Lightning Source LLC
Chambersburg PA
CBHW080257180526
45167CB00006B/2564